T0233990

INTERNATIONAL CENTRE FOR MECHANICAL SCIENCES

COURSES AND LECTURES - No. 279

SECURE DIGITAL COMMUNICATIONS

EDITED BY

G. LONGO
UNIVERSITA' DI TRIESTE

SPRINGER-VERLAG WIEN GMBH

This volume contains 118 illustrations.

This work is subject to copyright.
All rights are reserved,
whether the whole or part of the material is concerned
specifically those of translation, reprinting, re-use of illustrations,
broadcasting, reproduction by photocopying machine
or similar means, and storage in data banks.
© 1983 by Springer-Verlag Wien
Originally published by Springer Verlag Wien-New York in 1983

ISBN 978-3-211-81784-1 ISBN 978-3-7091-2640-0 (eBook)
DOI 10.1007/978-3-7091-2640-0

PREFACE

The necessity of keeping certain information secret is as old as human communication. At the same time the advantages to be gained in intercepting secret messages was soon realized. This has led to a century-long battle between the "codemakers" and the "codebreakers", the battle being fought in the everchanging arena of communication media.

The concept of secure communication itself has evolved through the ages, as the traditional communication systems – like the mail and courier services – were supplemented by more technical media – radio, telephone, television, data links, etc. The amount of protection needed also vary depending both on the importance of the message contents and on the determination of the interceptor to understand and use the information.

The most common forms of communication are highly insecure, since they are easily intercepted and understood. Probably the use of a courier is the safest way to exchange messages, but this is seldom possible in our present society, which is so much dependent on large amounts of information exchanged at very high rate. One possibility is then to make information (almost) non-interceptible by concealing the messages via a suitable transformation before transmission. A cipher system is precisely a more or less sophisticated way of doing this, and the "art" (some would prefer "science") of designing cipher systems is named cryptology (from two Greek words meaning "discourse on hiding").

It has to be emphasized that the revolution of microelectronics has led both to ever-increasing capabilities of and to ever increasing dependence upon data processing and data transmission systems. This results in a tremendous growth of the need for protection against unauthorized access and misuse of the data. While the general public is increasingly alarmed by frauds in this area, professional cryptologists are well aware of the weaknesses of present-day security and of the difficulties of their task.

The aim of this book is to give a broad survey of the problem of secure communication. Some contributions are introductory, some are more advanced, but the interested reader can benefit from reading all of them as different chapters on one single topic of increasing importance.

I wish to thank all the participants and all the lecturers, whose generous effort make it possible to publish this book.

Giuseppe Longo

Udine, November 1983.

CONTENTS

ELEMENTS OF CRYPTOLOGY.

M. Davio, J.-M. Goethals.
Philips Research Laboratory,
Ave. van Becelaere, 2, Box 8,
B 1170, Brussels.

1. CLASSICAL APPROACH AND SHANNON THEORY.

1.1. Definition of a cipher system.

This section is based on Shannon's original paper[1] which presents an information-theoretic approach to cryptology. Previous accounts of Shannon's theory may be found in the books by Konheim[2] and Beker and Piper[3] Figure 1 gives a schematic diagram of a cipher system (or secrecy system, as it was called by Shannon) At the transmitting end there are two "information" sources: a message source and a key source. Before any message is sent, the two parties, the encipherer and the recipient, agree on their key K, which is selected from the available set : the ***key space***. Once the key is agreed, the encipherer selects a message M from the ***message space***, enciphers it with the particular transformation T_K determined by the key, and sends the cryptogram $C = T_k(M)$ over a public channel (where it can be intercepted) to the intended recipient. At the receiving end the cryptogram and the key are combined by the decipherer to recover the message $M = T_K^{-1}(C)$. The set of all possible cryptograms is called the ***cryptogram space***. Naturally, the transformations T_k mapping messages into cryptograms should be invertible.

We will assume that each key K_i has an associated probability P_i and similarly that each message M_j has an associated probability q_j. Thus, following Shannon, we may define a Cipher System to consist of a ***finite family of reversible transformations from a set of messages, into a set of cryptograms***. There is a probability distribution of the set of possible messages and of the set of possible keys (or, equivalently, of the transformations).

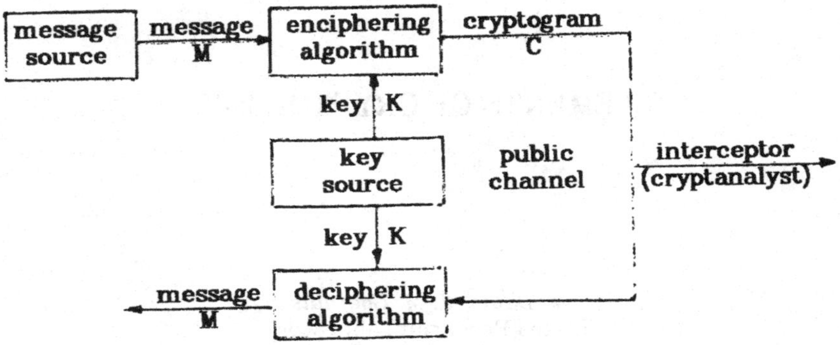

Figure 1.
Cipher system.

We assume that the cryptanalyst knows the system and the probability distributions. We shall mainly be concerned with the amount of secrecy that the system may provide, which is measured in terms of the uncertainty that the cryptanalyst has about the message or the key used for a specific cryptogram.

A cipher system may be conveniently represented as a bipartite graph with messages and cryptograms as vertices and transformations as edges. For each message M_i there will be exactly one edge emerging for each different key. Furthermore since each transformation is reversible, the same key cannot map distinct messages into the same cryptogram, and so each cryptogram has at most one edge entering it for each key. If there is exactly one edge entering each cryptogram for each key, then every possible cryptogram can be deciphered uniquely for each key and the cipher system is called *closed*.
Figure 2 gives an example of a closed system.

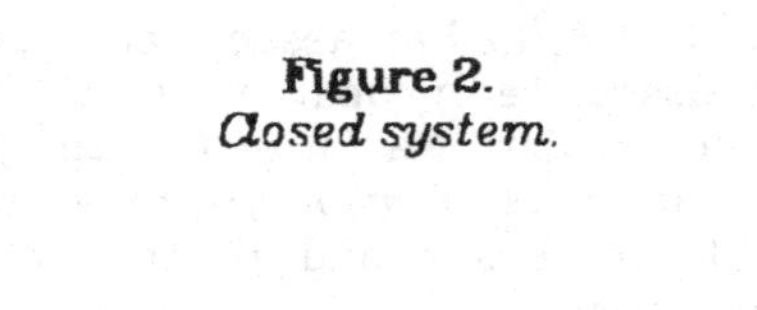

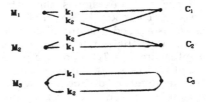

Figure 2.
Closed system.

1.2. Some examples

In this section, we shall briefly discuss some "classical" examples, where, unless otherwise stated, the messages and cryptograms consist of sequences of letters from the English alphabet. In practice the message space is restricted to those sequences which have some meaning in the language.

1.2.1. Simple substitutions (Monoalphabetic Cipher)

In a monoalphabetic cipher, each letter of the message is replaced by a fixed substitute, which is another letter of the alphabet. The key is thus a permutation of the alphabet. Denoting this permutation by f, we may write the cryptogram C obtained from the message

$$M = m_1 m_2 m_3 ....$$

as

$$C = f(m_1) f(m_2) f(m_3) \cdots$$

There are 26! possible permutations, each defined by a sequence of 26 letters: the permuted alphabet. (The Caesar Cipher belongs to that category. Here the permuted alphabet is obtained by cyclically shifting each letter three positions to the right).

1.2.2. Transpositions (with fixed period d).

The message is divided into groups of length d and a fixed permutation is applied to each group of d letters. The permutation is the key, which may thus be represented by a permuted sequence of the integers 1,2,...,d.

Ex. d=5, key = 2,3,1,5,4 (permutation (1,2,3)(4,5))

$$M = m_1 m_2 m_3 m_4 m_5 m_6 m_7 m_8 m_9 m_{10} \cdots$$

$$C = m_2 m_3 m_1 m_5 m_4 m_7 m_8 m_6 m_{10} m_9 \cdots$$

There are d! possible keys for a period d. (The Scytale cipher belongs to that category)

1.2.3. Polyalphabetic ciphers.

In these ciphers the key consists of a series of permuted alphabets. The first letter m_1 of the message is replaced by its substitute in the first alphabet, the second by its substitute in the second alphabet, and so on. Thus denoting by f_i the permutation induced by the i-th alphabet, we have

$$M = m_1 m_2 m_3 \cdots$$

$$C = f_1(m_1) f_2(m_2) f_3(m_3) \cdots$$

i.e.

$$c_i = f_i(m_i).$$

Naturally, for practical reasons, the number of permuted alphabets should be kept finite, in which case they are used repeatedly in a periodic fashion.
A popular version of this kind of cipher is the *Vigenere* cipher, where each f_i is a cyclic shift of the Caesar type, which can be described mathematically as follows:

$$f_i : m_i \to m_i + k_i \ mod\ 26$$

where the letters are viewed as numbers from 0 to 25 by use of the correspondence $A<\to 0, B<\to 1, C<\to 2, \cdots Z<\to 25$.
In this case the sequence of alphabets can be described by a sequence of letters. If the sequence is a meaningful text, we have the "*running key Vigenère*", where the key-text can be as long as the message. If the letters of the key are chosen at random and never repeat, we have the *Vernam system*, which for binary alphabets produces the *one-time pad*.

1.2.4. N-gram substitutions, matrix system.

Rather than substitute for letters we can substitute for groups of letters, say N-letter blocks. The cipher alphabet then consists of 26^N N-grams. One method of performing N-gram substitution is to operate on successive blocks of N letters with a matrix A having an inverse in modulo 26 arithmetic. As in the Caesar cipher the letters are assumed to be numbers from 0 to 25, making them elements of the ring Z_{26}. With an invertible matrix $A = (a_{ij})$, an N-gram cryptogram $C = (c_1, c_2, \ldots, c_n)$ is obtained from the message $M = (m_1, m_2, \ldots, m_N)$ by the linear transformation

$$c_i = \sum_j a_{ij} m_j \ (mod\ 26).$$

The matrix is the key and deciphering is performed with the inverse matrix.
After having examined some simple cipher systems we now study, in general terms, how to combine them.

1.3. Combining ciphers

Given two cipher systems $\mathbf{T} = \{T_K\}$ and $\mathbf{R} = \{R_K\}$ Shannon defined two ways of combining them to obtain a third systems $\mathbf{S} = \{S_K\}$. The first consists of combining them by taking a *weighted sum*

$$\mathbf{S} = \mathrm{p}\mathbf{T} + \mathrm{q}\mathbf{R}\ ,\ \mathrm{p}+\mathrm{q} = 1.$$

To encipher using **S** we must first choose between the systems **T** and **R**.
This choice which is part of the key is made with probabilities p and q, respectively. After this choice is made a particular transformation in **T** (or **R**) is selected with its original probability. Thus if $p_1, p_2, \ldots, p_n$ and $q_1, q_2, \ldots, q_n$ denote the respective probabilities of the transformations in **T** and **R**, then the m+n transformations in **S** have probabilities

$$pp_1, pp_2, \ldots, pp_n, qq_1, qq_2, \ldots, qq_n.$$

Obviously we may generalize this to weighted sums of an arbitrary number of cipher systems.

A second way of combining ciphers is by forming their *product*. Suppose **T** and **R** are two cipher systems such that the cryptogram space of **T** can be identified with the message space of **R**. Then we can first encipher using **T** and then *superencipher* the resulting cryptogram using **R**. The result is a cryptogram in the cryptogram space of **R**. Thus we have defined a new cipher system **S** called the product of **T** and **R**, which we write **S** = **RT**. The following

$$S_{k,k'}(M) := R_k(T_{k'}(M))$$

is a typical transformation in **S**. Clearly, if p and q are the probabilities of R_k and $T_{k'}$, respectively, then $S_{k,k'}$ has probability pq.

When, for a cipher system, the message and cryptogram spaces are equal, the system is called *endomorphic*. Clearly, if **T** and **R** are endomorphic with the same message and cryptogram spaces, the same holds for their product **S** = **RT**. Shannon showed that, with respect to the operations of taking weighted sums and products, *the set of* all *endomorphic ciphers* on a given space *is a linear associative algebra*.

1.4. Pure Ciphers

Now that we have defined a cipher system and studied a few examples we can begin to consider how "good" a system is. To do this we need to introduce some of Shannon's ideas, the first being the concept of a pure cipher. Clearly if we are going to investigate the effect of trying all possible transformations of a system, we must look at what happens when we try to decipher using the wrong one. Shannon defined a *pure cipher* to be one in which, whenever we encipher with one transformation, decipher with a second and then encipher again with a third, we can find a single transformation in the system which has the same effect.

Definition. A cipher system **T** is pure if, for every $T_i, T_j, T_k \in \mathbf{T}$ there is a $T_s \in \mathbf{T}$ such that

$$T_i T_j^{-1} T_k = T_s,$$

and every key is equally likely.

Theorem 3.1. In a pure cipher **T** the operations $T_i^{-1} T_j$ which transform the message space into itself form a group of permutations, whose order is equal to the number of keys.

Proof :

(i) Since for every i,j,k, there exists an s such that $T_j^{-1} T_k = T_i^{-1} T_s$, we may write

$$(T_h^{-1} T_i)(T_j^{-1} T_k) = T_k^{-1}(T_i T_i^{-1}) T_s = T_h^{-1} T_s,$$

which shows that the product of any two elements in the set is another element in the set.

(ii) For every i,j the inverse of $T_i^{-1} T_j$, belongs to the set. Hence this set, which we denote by G (**T**), is a group.

(iii) The order of this group is at least equal to the number of keys and at most equal to its square. Let

$$\mathbf{T} = T_1, T_2, \ldots, T_m$$

Then ∀i, the elements $T_i^{-1} T_j$, j=1,2,..., m are distinct, hence $|G(\mathbf{T})| \geq m$; and from part (i) of the proof we deduce $|G(\mathbf{T})| \leq m^2$. But we may express every $T_j^{-1}T_k$ in exactly m distinct ways as $T_i^{-1}T_s$. Hence $|G(\mathbf{t})| \leq m^2/m$, whence $|G(\mathbf{T})| = m$. ▪

This group G (**T**) partitions the message space into *equivalence classes* , called the *message residue classes*. The messages belonging to the same residue class are all the possible messages that can be obtained from anyone of them by first enciphering using any transformation T_j and then deciphering using the inverse T_i^{-1} of any other.

It is easy to see that the set G'(**T**) of the operations $T_j T_i^{-1}$ which transform the cryptogram space into itself is a group conjugate to G (**T**) and hence of the same order. In fact, for any transformation T_i, we have

$$G'(\mathbf{T}) = T_i \; G(\mathbf{T}) \; T_i^{-1}$$

The group G' (**T**) partitions the cryptogram space into equivalence classes, called the *cryptogram residue classes* . These are the images of the message residue classes under any transformation T_i in the system.

From classical results on permutation groups we have the following theorem, where we denote by $M_1, M_2, \ldots, M_s$ the message residue classes and by $C_1, C_2, \ldots, C_s$ the corresponding cryptogram residue classes.

Theorem 3.2. For the message and cryptogram residue classes of a pure system with m keys, the following holds:

(i) $|M_i| = |C_i| = m/g_i$, where g_i is the number of elements in G (**T**) which fix (= transform into itself) any given element in M_i.

(ii) for $m_i \in M_i$ and $c_j \in C_i$, there are exactly g_i transformations $T \in \mathbf{T}$ such that $T(m_i) = c_j$ Figure 3 gives an example of a pure cipher system.

1.5. Perfect Secrecy.

We shall now begin to consider the question of how much security a system offers, when a cryptanalyst has unlimited time and computational power available for the analysis of cryptograms. We thus assume that the cryptanalyst is able to use his knowledge of the system in use and of the probability distributions on the message and key spaces in his attempts to "solve" or decrypt one or several intercepted cryptograms.

Let us assume that we have a cipher system **T** with a finite message space $\mathbf{M} = \{m_1, m_2, \ldots, m_n\}$, a finite cryptogram space $\mathbf{C} = \{c_1, c_2, \ldots, c_m\}$ and transformations $T_1, T_2, \ldots, T_p$. Let us denote by p (m_i) the a priori probability of message m_i. After having intercepted the cryptogram C_j the cryptanalyst can calculate, at least in principle, the a posteriori probabilities of various messages, which we denote by $p(m_i \mid c_j)$. Shannon defined perfect secrecy by the condition that, for all c_j, the a posteriori probabilities are equal to the a priori probabilities, i.e. for i=1,2,...,n,

$$p(m_i \mid c_j) = p(m_i), \; \forall c_j.$$

In this case, intercepting the cryptogram has given the cryptanalyst no information. (The average over all possible messages and cryptograms of the quantity log $[p(m_i \mid c_j)/p(m_i)]$ measures the amount of information on m_i gained by the interception of c_j).

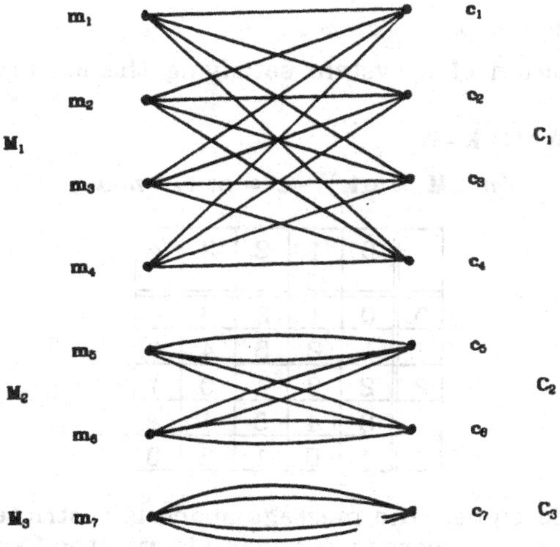

Figure 3
A pure cipher system

Since, by Bayes' theorem, we have

$$p(m_i)\,p(c_j\,|m_i) = p(c_j)\,p(m_i\,|c_j),$$

where $p(c_j)$ denotes the probability of obtaining c_j (from any message) and $p(c_j\,|m_i)$ that of obtaining it from m_i, we have the following theorem:

Theorem 4.1. A necessary and sufficient condition for perfect secrecy is that

$$p(c\,|m) = p(c)$$

for all messages m and cryptograms c. In other words, $p(c\,|m)$ must be independent of m.

Stated another way, perfect secrecy means that, for any messages m_i, m_j and any cryptogram C_k , the total probability of all keys which transform m_i into c_k is the same as that of all keys which transform m_j into c_k .

Since this probability is nonzero, we must have at least one key which transforms any m_i into any C_k. Hence ***the number of keys is at least equal to the number of distinct messages.*** (More precisely, it is at least equal to the number of distinct cryptograms which itself is at least equal to the number of messages, i.e. with the notations as above, we must have $p \geq m \geq n$).

Perfect systems in which these numbers are equal are characterized as follows:

Theorem 4.2. A cipher system in which the message, cryptogram, and key spaces have equal size has perfect secrecy if and only if:

(i) There is exactly one key transforming each message into each cryptogram, and

(ii) all keys are equally likely.

The matrix representation of a system satisfying the above condition is a *Latin square*.

Example. $m=n=p=5$, $\mathbf{M}=\mathbf{C}=\mathbf{K}=\mathbf{Z}_5$

$$(m \in \mathbf{M},\ k \in \mathbf{k}) \to c = m+k \ mod\ 5$$

	0	1	2	3	4
0	0	1	2	3	4
1	1	2	3	4	0
2	2	3	4	0	1
3	3	4	0	1	2
4	4	0	1	2	3

If, in a monoalphabetic cipher, the message space is restricted to one letter then clearly we can attain perfect secrecy, for example with the Caesar cipher. But, if we increase the size of the message space to include longer words, without increasing the size of the key space, the security offered by the system falls rapidly. This is because the amount of uncertainty we can introduce in the system cannot be greater than the key uncertainty. Thus if we increase the size of the message space we should simultaneously increase the size of the key space. *If we want to conceal our messages completely we have to introduce uncertainty with the key at at least the same rate as we introduce information in our messages* (The Vernam system (or one-time pad) is unbeakable just because it obeys the above rule).

1.6. Random Ciphers

In this section we shall mainly be concerned with the amount of cryptogram needed before a unique solution can be obtained for the key. Shannon called it the *unicity distance*. The concepts of redundancy, entropy and equivocation play a central role in this discussion.

1.6.1. Entropy and equivocation

In a cipher system there are two statistical choices involved: those of the message and the key. The message and key entropies

$$H(M) = -\sum p(m) \log p(m)$$

$$H(K) = -\sum p(k) \log p(k)$$

measure the amount of uncertainty involved in these choices.

When a cryptanalyst has intercepted some ciphertext there is still some uncertainty about the message or the key used. This uncertainty is measured by the equivocations (or conditional entropies)

$$H(M \mid C) = - \sum_{m,c} p(c) p(m \mid c) \log p(m \mid c),$$

$$H(K|C) = -\sum_{k,c} p(c)p(k|c)\log p(k|c).$$

The differences $H(M)-H(M|C), H(K)-H(K|C)$ measure the amount of information gained by intercepting a cryptogram. Naturally these quantities vary with the length N of messages and intercepted cryptograms. (Here we assume the key has a fixed length). We shall use the notations $H(M^N|C^N), H(K|C^N)$ when the messages and cryptograms involved have length N. Shannon proved the following ***properties*** :

Theorem 5.1. The equivocation of key is a non-increasing function of N.

The equivocation of the first A Letter of a message of a non-increasing function of the number N of intercepted letters. For a given number of intercepted letters, the equivocation of the message is less than or equal to that of the key. This may be written:

(i)

$$H(K|C^{N+1}) \leq H(K|C^N)$$

(ii)

$$H(M^A|C^{N+1}) \leq H(M^A|C^N)$$

(iii)

$$H(M^N|C^N) \leq H(K|C^N)$$

The last inequality follows from

$$\begin{aligned} H(M,K|C) &= H(M|C) + H(K|M,C) \\ &= H(K|C) + H(M|K,C), \end{aligned}$$

where $H(M|K,C)$ is zero since (K,C) determine M uniquely, but $H(K|M,C)$ generally is not.

Considering the joint entropy $H(M,K)$, we may write

$$H(M,K) = H(M) + H(K) \tag{1}$$

since M and k are statistically independent. On the other hand, since $(M,K)\rightarrow C$ or $(C,K)\rightarrow M$, we have

$$\begin{aligned} H(M,C,K) &= H(M,K) = H(C,K), \\ H(C,K) &= H(C) + H(K|C). \end{aligned} \tag{2}$$

Hence, for the equivocation $H(K|C)$, we may write

$$H(K|C) = H(M) + H(K) - H(C),$$

or equivalently, (now specifying the length N).

$$H(K|C^N) = H(K) - (H(C^N) - H(M^N)) \tag{3}$$

Let us assume that M and C use the same alphabet A. Let $R_0 = \log A$. Then,

$$H(C^N) \leq N\,R_0.$$

On the other hand, for N sufficiently large, $H(M^N)$ is approximately equal to

$$H(M^N) \approx NR$$

where R is the average entropy per letter in the language used. From (3) we deduce

$$H(K \mid C^N) \geq H(K) - N(R_0 - R)$$

The quantity $R_0 - R$ is the *redundancy* D per letter in the language. The above inequality may be written

$$H(K) - H(K \mid C^N) \leq ND$$

which shows that the amount of uncertainty about K removed by the interception of N letters of cryptograms is not greater than the redundancy of N letters of messages.

1.6.2. Definition of a random cipher

Shannon defined a random cipher to be a cipher system satisfying the following three conditions:

(i) The number of possible messages of length N is

$$T = 2^{NR_0}.$$

(ii) The possible messages of length N can be divided into two groups: one group of high and fairly uniform a priori probability (the meaningful messages), and a second group of negligibly small total probability. The high probability group contains

$$S = 2^{NR}$$

messages, with $R = (1/N)H(M^N)$ is the entropy per letter for N-letter messages.

(iii) There are k equiprobable keys and for the random cipher, we assume that in the bipartite graph representing the cipher, the k edges from each cryptogram go back to a random selection of the possible messages.

Actually a random cipher is an ensemble of ciphers and the equivocation is the average for this ensemble.

1.6.3. Redundancy and unicity distance

The quantity $D = R_0 - R$ represents the amount of redundancy per letter in N-letter messages. For the random cipher, for any given cryptogram, the probability of obtaining a meaningful decipherment with any given key is

$$S/T = 2^{-N(R_0 - R)} = 2^{-ND}.$$

Since there are $k = 2^{H(K)}$ equiprobable keys, the *expected* value of the *number of meaningful decipherments* is

$$2^{H(K) - ND}$$

If $H(K) \gg ND$, then there is a great chance of obtaining a meaningful decipherment and thus a low likelihood of determining the correct key and message.

If, on the other hand $H(K) << ND$, then as soon as a meaningful decipherment is obtained it has a fair chance of being the correct one. The borderline between these two situations is determined by the quantity

$$N_0 = H(K)/D$$

of intercepted ciphertext, which was called by Shannon, the *unicity distance*. For any amount of ciphertext exceeding that value there is a fair chance of determining the key uniquely.

References

1. C. E. Shannon, "Communication theory of secrecy systems," *BSTJ* **28** pp. 656-715 (1949).
2. A. G. Konheim, *Cryptography: A Primer*, J Wiley, New York (1981).
3. H. Beker and F. Piper, *Cipher systems*, Northwood Books, London (1982).

2. BLOCK CIPHERS

2.1. Definitions.

A *block cipher* is a system of transformations of blocks of n bits into blocks of n bits. Since these transformations should be invertible, they act as ***permutations*** on the set $V_n = \{0,1\}^n$ or, equivalently, on the set of 2^n numbers $\{0,1,\dots, 2^n - 1\}$.

$$\mathbf{x} = (x_0, x_1, \dots, x_{n-1}) \leftrightarrow x = \sum_{i=0}^{n-1} x_i 2^i$$

Thus, each transformation is an element of $SYM(2^n)$, the symmetric group of all of the permutations of degree 2^n.

The selection of a particular transformation is determinated by the *key*, which generally consists of a finite sequence of bits. If every possible permutation of $SYM(2^n)$ would be included in the Cipher, we would need

$$\log_2(2^n!) \approx n.\, 2^n \textit{ bits},$$

a prohibitively large number.

Besides, it would be extremely difficult to design a procedure which would determine the permutation from the key.

Thus, in general, a block cipher consists of a smaller set of permutations which are obtained by combining certain simple operations. The way they are combined is determined by the key.

We shall first study some of these simple operations which can be used as ***building blocks***. These are:

- bit permutations (wire crossing)
- translations
- linear transformations
- addition (with carry) mod 2^n
- substitutions (S - boxes)

We shall then briefly discuss how these operations can be combined. We then conclude with a short discussion of their security.

2.2. Requirements on a good block cipher.

But, before doing that, let us briefly examine *what we do require of a good block cipher*.

A block cipher can be viewed as a substitution cipher on the alphabet of n - bit words, and therefore can be vulnerable to cryptanalytic attacks based on statistical analysis.

These attacks can be thwarted by making the *value of n sufficiently large* so that the alphabet size 2^n exceeds traitable values.

Moreover, we should also require the *key size* to be *large enough* to prevent a key trial.

Finally, the *key* should be *involved in* a sufficiently *complex way* in the enciphering process so that no simple method, other than key trial, would be possible of recovering the key from the knowledge of matched plaintext - ciphertext pairs.

We should bear this in mind when examining possible solutions.

2.3. Building blocks.

(i) *bit permutations*: consist of permuting the n bit positions, which can be obtained by simple wire crossing.

(ii) *translations*: (bit-by-bit modulo 2 addition)

$$\mathbf{m} \rightarrow \mathbf{m} \oplus \mathbf{k} = \mathbf{c}$$

$$(\text{matched pair } (\mathbf{m},\mathbf{c}) \rightarrow \mathbf{k} = \mathbf{m} \oplus \mathbf{c})$$

(iii) *linear transformations*: (with A non singular n log n matrix over GF (2))

$$\mathbf{m} \rightarrow A\mathbf{m} = \mathbf{c}$$

(with n linearly independent m_i and corresponding c_i, we may determine A)

m_0	m_1	m_2	c_0	c_1	c_2
0	0	0	0	1	0
1	0	0	0	1	1
0	1	0	1	0	0
1	1	0	0	0	0
0	0	1	1	1	1
1	0	1	0	0	1
0	1	1	1	0	1
1	1	1	1	1	0

Figure 1.

(iv) *addition mod* 2^n. (Here we consider m to be a number in the set $\{0,1,\ldots,2^n-1\}$)

$$m \rightarrow m + k \bmod 2^n = c$$

$$(c - m \ mod\ 2^n = k)$$

(v) *substitutions* (S - boxes)

ex: the permutation $\pi = (0,2,1,6,5,4,7,3)$ corresponds to the truth table given in figure 1.

Each c_i is a Boolean function of (m_0, m_1, m_2). If the number of variables is not too large, such functions can easily be implemented (for example, by table look-up).

Since, for a good block cipher, n should be large, we are more or less forced to divide the n -bit block into smaller blocks on each of which we may define a substitution as above.

2.4. Combining building blocks.

Individually, each of the above building blocks is a poor cipher. However, by combining them (as suggested by Shannon) in a product cipher, we may obtain much stronger ciphers.

Some of the combinations, however, are weak. For example, by combining translations and linear transformations we obtain an ***affine cipher***:

$$\mathbf{m} \to A\mathbf{m} + \mathbf{b}$$

which can be broken in much the same way as a linear cipher. Thus affine substitutions should be avoided.

A stronger combination is obtained with translations and mod 2^n addition, a system which was studied by Edna Grossman, [1] who showed that the group generated by these two types of addition has order:

$$2^{(2^{n-1} + n - 1)}$$

(The problem of generating the full symmetric group *SYM* (2^n) with simple transformations was also studied by E. Grossman and D. Coppersmith.)[2]

The ciphers designed at IBM mostly use a combination of bit permutations (P - boxes) and substitutions on smaller blocks (S - boxes). In particular the Feistel cipher from which the DES was derived is of that type.

References

1. E. Grossman, "Group theoretic remarks on cryptogtaphic systems based on two types of addition," IBM TJ Wattson Res. Center RC 4742 (1974).

2. D. Coppersmith and E. Grossman, "Generators for certain alternating groups with applications to cryptography," *SIAM journal on applied mathematics* **29** pp. 624-627 (1975).

3. SUBSTITUTION NETWORKS

3.1. Introduction

Substitution networks play a central rôle in cryptology. They indeed define bijections on an alphabet that will play the roles of *cleartext* and of *ciphertext* alphabet. The essential feature of substitution networks is that this common alphabet is expressed as a product space. A substitution network may thus be viewed as a black-box with say n inputs and n outputs carrying multiplevalued input and output variables: each of the output variables will thus be viewed as a discrete function of the input variables. As, in practice, the handled variables will be binary, the outputs of a substitution network appear as Boolean functions of the binary inputs and the permutation realized by the network acts on the space of binary n-tuples.

In the present study, we shall first relate various fields where the concept of substitution network appears in a more or less explicit form. These fields are:

(i) the theory of substitution networks, as it is usually tackled in cryptology;[1,2,3]

(ii) the theory of permutation networks, initiated in telephony switching;[4,5,6]

(iii) the theory of shift registers, a branch of switching theory[7]

We shall in particular derive a canonical form of substitution networks which has the property of requiring the minimum possible number of control variables.

3.2. Definitions

A ***substitution network*** on the alphabet Σ ($|\Sigma| = N$) is sketched in figure 1.

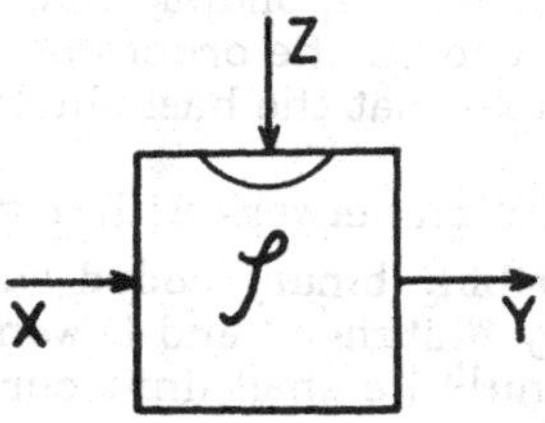

Figure 1.
Substitution network.

This network has:

(i) one N-valued input X;
(ii) one N-valued output Y;
(iii) one M-valued control input Z.

For each value of the control input Z, the network realizes a bijection:

$$f_Z : \Sigma \to \Sigma \tag{1}$$

on the alphabet Σ. Each of the bijections f_Z is a ***substitution***. The set $S = \{f_Z \mid Z \in \mathbf{Z}_M\}$ * thus completely characterizes the substitution network. When the network is able to realize the N! possible substitutions on Σ, it is called ***total substitution network***. In a total substitution network, the condition:

$$M \geq N! \tag{2}$$

should obviously be satisfied, as there should be at least one control setting for each of the substitutions to be realized.

The importance of substitutions for cryptographic applications was already clearly perceived in the now classical paper of Shannon[8] Since then, the concept of substitution network has been paid additional attention[9, 10, 1, 2] Popular examples of substitutions are provided by the well known Caesar and Vigenere ciphers. In our model, X then becomes a ***cleartext literal***, Y is a ***ciphertext literal***, and the bijective nature of the functions f_Z allows one to recover the plaintext from the ciphertext.

When the size of Σ becomes large, the input literals, the output literals and the keys (i.e. the values of the control input) will be encoded as vectors of m-valued variables. In this case, we write $\mathbf{x}$, $\mathbf{y}$ and $\mathbf{z}$ instead of X, Y and Z, respectively and we have:

$$\begin{aligned}\mathbf{x} &= [x_{n-1},\ldots,x_1,x_0],\\ \mathbf{y} &= [y_{n-1},\ldots,y_1,y_0],\\ \mathbf{z} &= [z_{m-1},\ldots,z_1,z_0].\end{aligned} \tag{3}$$

with $x_i, y_j, z_k \in \mathbf{Z}_r$. We should thus have $N = r^n$, $M = r^m$. In practice, we shall have $r = 2$, but for theoretical purposes, the general parameter r will often prove useful.

In the application of substitution networks to cryptology, one is immediately faced with two basic problems: the first problem is simply to describe or define a substitution network S; the second problem is, given a description of a substitution network S ,to obtain a description of the inverse substitution network S^{-1}, which, given $\mathbf{y}$ allows one to recover the corresponding input $\mathbf{x}$. These problems are by no means trivial if one takes into account the orders of magnitude adopted in modern cryptology; remember for example that the basic building block of the *Data Encryption Standard*
(DES), is a binary coded substitution network with $n = 64$ and $m = 56$.

General methods for describing binary coded substitution networks and their inverses have been described by Whitehead and Lowenheim[11, 12] An account of these results may be found in Rudeanu[13] We shall limit ourselves in this context to illustrate these ideas by an example.

*The symbol $\mathbf{Z}_M$ represents the set $\{0,1,\ldots(M-1)\}$ of integers mod M.

Example.

Figure 2 defines by its truth table a binary substitution network.

x_2	x_1	x_0	y_2	y_1	y_0
0	0	0	0	1	1
0	0	1	0	0	0
0	1	0	1	0	1
0	1	1	1	0	0
1	0	0	1	1	0
1	0	1	0	1	0
1	1	0	1	1	1
1	1	1	0	0	1

Figure 2.
Truth table of a substitution network.

The network is described by the Boolean equations:

$$\begin{aligned} y_2 &= \bar{x}_2 x_1 \bar{x}_0 \cup \bar{x}_2 x_1 x_0 \cup x_2 \bar{x}_1 \bar{x}_0 \cup x_2 x_1 \bar{x}_0 \\ y_1 &= \bar{x}_2 \bar{x}_1 \bar{x}_0 \cup x_2 \bar{x}_1 \bar{x}_0 \cup x_2 \bar{x}_1 x_0 \cup x_2 x_1 \bar{x}_0 \\ y_0 &= \bar{x}_2 \bar{x}_1 \bar{x}_0 \cup \bar{x}_2 x_1 \bar{x}_0 \cup x_2 x_1 \bar{x}_0 \cup x_2 x_1 x_0 . \end{aligned} \tag{4}$$

The observation of the truth table immediately allows one to obtain a Boolean description of the inverse substitution network. One has indeed:

$$\begin{aligned} x_2 &= y_2 y_1 \bar{y}_0 \cup \bar{y}_2 y_1 \bar{y}_0 \cup y_2 y_1 y_0 \cup \bar{y}_2 \bar{y}_1 y_0 \\ x_1 &= y_2 \bar{y}_1 y_0 \cup y_2 \bar{y}_1 \bar{y}_0 \cup y_2 y_1 y_0 \cup \bar{y}_2 \bar{y}_1 y_0 \\ x_0 &= \bar{y}_2 \bar{y}_1 \bar{y}_0 \cup y_2 \bar{y}_1 \bar{y}_0 \cup \bar{y}_2 y_1 \bar{y}_0 \cup \bar{y}_2 \bar{y}_1 y_0 . \end{aligned} \tag{5}$$

The crucial fact to be observed about binary substitution networks is that the $\mathbf{x}$ and $\mathbf{y}$ minterm sets are related by a permutation matrix. Thus, if we denote by $\mathbf{m}(\mathbf{x})$ the column matrix of the 2^n minterms in $\mathbf{x}$ (ordered by lexicographic order, we shall have:

$$\mathbf{m}(\mathbf{y}) = \mathbf{P}\,\mathbf{m}(\mathbf{x}). \tag{6}$$

If we denote by an upper T the matrix transposition, we immediately obtain the description of the inverse substitution network:

$$\mathbf{m}(\mathbf{x}) = \mathbf{P}^T \mathbf{m}(\mathbf{y}). \tag{7}$$

While extremely simple, the above example and equations should not hide the inherent complexity of our two basic problems. There are indeed $(2^n)!$ substitutions on the set of binary n-tuples; the very definition of an arbitrary substitution on this set thus requires $O(n2^n)$ bits, and this shows that the description of an arbitrary substitution network by its truth table is asymptotically optimal. Clearly enough, the amount of information required by such a definition is prohibitive for practical values of n and this will orient us to the study of particular classes of substitution networks.

3.3. Huffman's theorem

An interesting characterization of substitution networks has been given by Huffman[14] In the present section, we shall slightly strengthen that property. We first introduce some definitions and notation. We consider a substitution on $\{0,1\}^n$:

$$S : \{0,1\}^n \to \{0,1\}^n : \mathbf{x} \to \mathbf{y} = S(\mathbf{x}),$$

for example, the substitution network defined in figure 2. Such a network is thus defined by n Boolean functions of n variables. Let us recall that any Boolean function $f(\mathbf{x})$ of n variables is described by a vector having 2^n binary components called ***truth vector*** of the function. The ***weight*** $w(f)$ of the function f is the number of components with value 1 in the truth vector of f.

Consider now a subset

$$T \subseteq \mathbf{Z}_n : |T| = t$$

and define the function

$$Y(T) = \prod_{i \in T} y_i .$$

Theorem 1. *The function S is a substitution on $\{0,1\}^n$ if and only if the two following conditions are satisfied:*

$$\forall T \subseteq \mathbf{Z}_n : |T| = n-1 : w(Y(T)) \text{ is even};$$

$$w(Y(\mathbf{Z}_n)) \text{ is odd}$$

Proof. The necessity is obvious. Indeed, if one considers the truth table defining the functions y_j, one immediately observe that if $|T| = t$, one has: $w(Y(T)) = 2^{n-t}$.
The sufficiency is almost as easy. By hypothesis, we have, for example:

$$w(y_{n-2} \cdots y_1 y_0) \text{ is even};$$

$$w(y_{n-1} y_{n-2} \cdots y_1 y_0) \text{ is odd}.$$

These assertions immediately imply:

$$w(\bar{y}_{n-1} y_{n-2} \cdots y_1 y_0) \text{ is odd}.$$

Complementing the y_i's in all the possible ways, one then reaches the conclusion:

$$w(\mathbf{m}(\mathbf{y})) \text{ is odd}$$

for every minterm $\mathbf{m}(\mathbf{y})$, and this implies, as there are exactly n disjoint such minterms:

$$w(\mathbf{m}(\mathbf{y})) = 1$$

for every minterm. This assertion in turn implies that each of the 2^n possible binary patterns appear once and only once in the truth table of S.

■

3.4. Golomb's theorem

In the present section, we relate the concept of substitution network to some aspects of the shift-register theory. We consider a substitution:

$$f : \Sigma \longrightarrow \Sigma, \tag{8}$$

and we assume that the alphabet Σ is the product space:

$$\Sigma = \Sigma_1 \times \Sigma_0 \tag{9}$$

with:

$$|\Sigma_1| = N_1;\ |\Sigma_0| = N_0;\ |\Sigma| = N = N_1 N_0. \tag{10}$$

We furthermore assume, without loss of generality that:

$$\Sigma_1 = \mathbf{Z}_{N_1};\ \Sigma_0 = \mathbf{Z}_{N_0}.$$

An element X of Σ may thus be represented by its coordinates (x_1, x_0) with $x_1 \in \Sigma_1$ and $x_0 \in \Sigma_0$. The coordinates x_1 and x_0 are referred to as the *block coordinate* and the *point coordinate* , respectively.

To introduce the discussion, let us consider the network shown in figure 3 and let us derive the conditions to be fulfilled in order that this network were a substitution network.

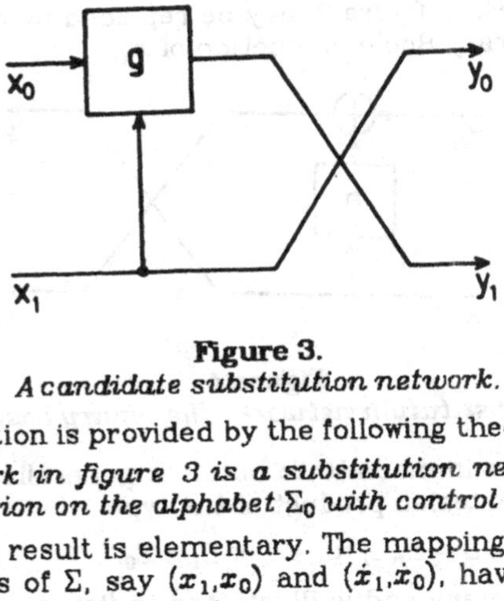

Figure 3.
A candidate substitution network.

The answer to that question is provided by the following theorem.

Theorem 2. *The network in figure 3 is a substitution network if and only if the function g is a substitution on the alphabet Σ_0 with control variable x_1.*

Proof. The proof of this result is elementary. The mapping $f : \Sigma \to \Sigma$ is a substitution iff two distinct elements of Σ, say (x_1, x_0) and $(\dot{x}_1, \dot{x}_0)$, have distinct images (y_1, y_0) and $(\dot{y}_1, \dot{y}_0)$ by f. Now, two cases may arise:

(i) $x_1 \neq \dot{x}_1$. In that case, we have by construction: $y_0 \neq \dot{y}_0$, and there are no additional conditions.

(ii) $x_1 = \dot{x}_1$ and $x_0 \neq \dot{x}_0$. In this case, we have $y_0 = \dot{y}_0$ and we expect $y_1 \neq \dot{y}_1$; to summarize, we should have:

$$\forall x_1 \in \Sigma_1 : x_0 \neq \dot{x}_0 \longrightarrow y_1 \neq \dot{y}_1. \tag{11}$$

and the latter condition actually imposes to the function g to be itself a substitution.

Comments.

(i) The output wire crossing in figure 3 is clearly not essential in the proof of theorem 2. Its interest may intuitively be understood as follows: the substitution carried out by the network in figure 3 exchanges between themselves points belonging to the same block. To be able to generate a total substitution network, we should also be able to exchange between themselves points belonging to distinct blocks. A possible way to reach this goal could be to exchange blocks between themselves, while keeping constant the point coordinate. As a matter of fact, the output wirecrossing in figure 3 allows one to carry out the latter operation by a second cell of the same type.

(ii) An interesting particular case of the network under study arises when $\Sigma_0 = \{0,1\}$. In that case, the implication (11) becomes:

$$\forall x_1 \in \Sigma_1 : x_0 = \bar{x}_0 \longrightarrow y_1 = \bar{y}_1, \tag{12}$$

and the most general way of satisfying (12) is obviously to adopt:

$$y_1 = x_0 \oplus h(x_1). \tag{13}$$

where O denotes the mod two sum. This is the essence of a theorem by Golomb[7] In this case, the network in figure 3 may be replaced by the network in figure 4 where $g(x_1)$ is an arbitrary Boolean function of x_1.

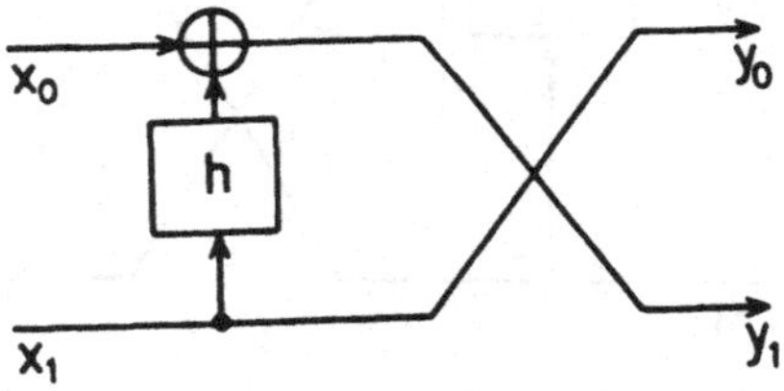

Figure 4.
Substitution network. The binary case.

(iii) Observe finally that it is immediate to obtain a generalization of theorem 2 to the case where Σ is a Cartesian product of the type:

$$\Sigma = \Sigma_{n-1} \times \cdots \times \Sigma_1 \times \Sigma_0.$$

The extension is elementary and is illustrated in figure 5 for $n=3$. The affinity of our basic cell with a shift register should now be clear: indeed, if the combinational cell displayed in figure 5 is completed by a register and if the outputs of that register are fed back to the cell inputs, one obtains the scheme of an autonomous shift register in the sense of Golomb.

3.5. Canonical form of substitution networks

We again consider a substitution:

$$f : \Sigma_1 \times \Sigma_0 \longrightarrow \Sigma_1 \times \Sigma_0$$

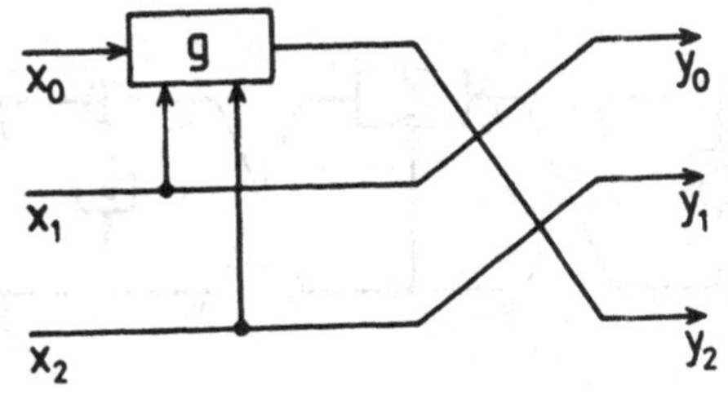

Figure 5.
Substitution networks and shift registers.

on some product space. The basic result to be described in this section is easily understood in the light of figure 6 and it is expressed by the following theorem.

Theorem 3. *Any substitution on the product space $\Sigma_1 \times \Sigma_0$ may be decomposed as the product of a point substitution, of a block substitution and of a point substitution.*

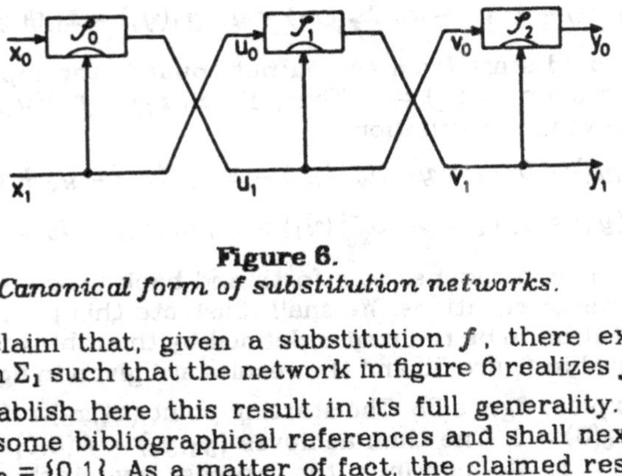

Figure 6.
Canonical form of substitution networks.

More precisely, we claim that, given a substitution f, there exist substitutions S_0, S_2 on Σ_0 and S_1 on Σ_1 such that the network in figure 6 realizes f.

We shall not establish here this result in its full generality. We shall merely orient the reader to some bibliographical references and shall next concentrate on the particular case $\Sigma_0 = \{0,1\}$. As a matter of fact, the claimed result is classical in the context of telephony switching where it is known as the Slepian-Duguid theorem[5, 15] A complete account of the proof may be found in Benes[4] Practical algorithms for computing S_0, S_1, S_2 may be found in Neiman[16] and Tsao-Wu[17] The above methodology was suggested by Ronse[3]

While constructive, the methods we have just mentioned are not completely satisfactory from a computational point of view. The determination of S_0, S_1 and S_2 has indeed a high computational complexity. The involved computations are however much simpler in the particular case $\Sigma_0 = \{0,1\}$. In this case, a much simpler computation algorithm, due to Waksman[6] and known as the *looping algorithm* is available. We shall briefly reformulate this algorithm in substitution network terms.

Let us first observe that, whenever $\Sigma_0 = \{0,1\}$, the general network in figure 6 may be replaced by that in figure 7. Basically, we have to determine

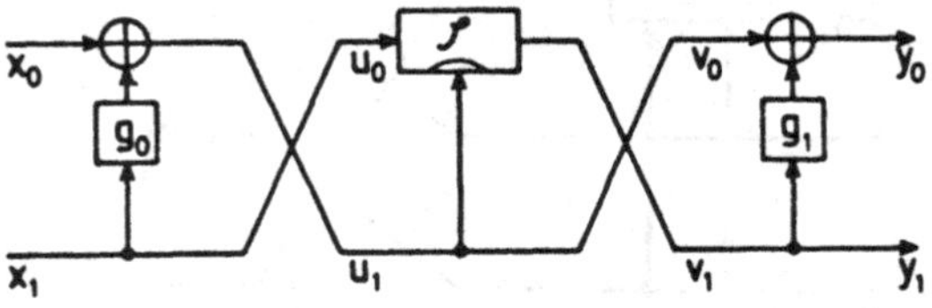

Figure 7.
Canonical substitution network: the binary case.

- the binary functions $g_0(x_1)$ and $g_1(v_1)$ defined on Σ_1;
- two substitutions $S_{u_1}(u_0), u_1 \in \{0,1\}$ on Σ_1.

The crucial point in the understanding of the looping algorithm may be understood as follows: let $(y_1, y_0) = f(x_1, x_0)$. There are basically two possibilities of choosing g_0, g_1 and S to satisfy that requirement, viz.:

$$g_0(x_1) = 0;\ u_1 = x_0;\ S_{x_0}(x_1) = y_1;\ g_1(y_1) = x_0 \oplus y_0.$$

$$g_0(x_1) = 1;\ u_1 = \bar{x}_0;\ S_{\bar{x}_0}(x_1) = y_1;\ g_1(y_1) = \bar{x}_0 \oplus y_0.$$

Symmetrically, we could start from the output towards the input by either one of the choices $g_1(y_1) = 0$ or $g_1(y_1) = 1$. Thus, if $(x_1, x_0) = f^{-1}(y_1, y_0)$, we shall have either one of the following two situations:

$$g_1(y_1) = 0;\ v_0 = y_0;\ S_{y_0}^{-1}(y_1) = x_1;\ g_0(x_1) = y_0 \oplus x_0.$$

$$g_1(y_1) = 1;\ v_0 = \bar{y}_0;\ S_{\bar{y}_0}^{-1}(y_1) = x_1;\ g_0(x_1) = \bar{y}_0 \oplus x_0.$$

The looping algorithm then reduces to a forth and back process using alternatively the two above systems of equations. We shall illustrate this process by a numerical example. The substitution to be realized is defined by the table in figure 8a; the computations are gathered in figure 8b and the results are given in figure 8.c

Let us briefly explain figure 8b. The starting point $(x_1, x_0) = (0,0)$ is arbitrary, as well as the choice $g_0(0) = 0$. One then achieves $(y_1, y_0) = f(0,0) = (2,1)$ by choosing $S_0(0) = 2$ and $g_1(2) = 1$. We continue the process by starting backwards from $(y_1, y_0) = (2,0)$ as this point belongs to the same output block as the output point $(2,1)$ reached in the previous step. To reach $(x_1, x_0) = f^{-1}(2,0) = (1,1)$, we have now to select: $S_1^{-1}(2) = 1$ or $S_1(1) = 2$ and $g_0(1) = 0$. The forth and back process is then continued until one reaches $(x_1, x_0) = (0,1)$, paired with the original starting point. The process is then restarted from any remaining input point (x_1, x_0).

It is not difficult, from the above discussion, to design a total substitution network on the alphabet $\Sigma = \Sigma_1 \times \Sigma_0$: clearly, g_0 and g_1 should be replaced by *Universal logical modules* (ULM's) able to realize arbitrary Boolean functions defined on Σ_1. Similarly, the network S should be able to realize an arbitrary substitution on Σ_1. This is illustrated in figure 9a which represents a total substitution network on $\Sigma = \{0,1\}^2$. This network uses three exclusive-OR gates and three multiplexers acting as ULM's on $\{0,1\}$. The behaviour of such a multiplexer is defined in figure 9b.

x	x_1	x_0	y	y_1	y_0
0	0	0	5	2	1
1	0	1	7	3	1
2	1	0	6	3	0
3	1	1	4	2	0
4	2	0	1	0	1
5	2	1	3	1	1
6	3	0	2	1	0
7	3	1	0	0	0

a)

$(x_1,x_0) = (0,0);\ g_0(0) = 0;\ S_0(0) = 2;\ g_1(2) = 1;\ (y_1,y_0) = (2,1)$

$(x_1,x_0) = (1,1);\ g_0(1) = 0;\ S_1^{-1}(2) = 1;\ g_1(2) = 1;\ (y_1,y_0) = (2,0)$

$(x_1,x_0) = (1,0);\ g_0(1) = 0;\ S_0(1) = 3;\ g_1(3) = 0;\ (y_1,y_0 = (3,0)$

$(x_1,x_0) = (0,1);\ g_0(0) = 0;\ S_1^{-1}(3) = 0;\ g_1(3) = 0;\ (y_1,y_0) = (3,1)$

$(x_1,x_0) = (2,0);\ g_0(2) = 0;\ S_0(2) = 0;\ g_1(0) = 1;\ (y_1,y_0) = (0,1)$

$(x_1,x_0) = (3,1);\ g_0(3) = 0;\ S_1^{-1}(0) = 3;\ g_1(0) = 1;\ (y_1,y_0) = (0,0)$

$(x_1,x_0) = (3,0);\ g_0(3) = 0;\ S_0(3) = 1;\ g_1(1) = 0;\ (y_1,y_0) = (1,0)$

$(x_1,x_0) = (2,1);\ g_0(2) = 0;\ S_1^{-1}(1) = 2;\ g_1(1) = 0;\ (y_1,y_0) = (1,1).$

b)

x_1	$g_0(x_1)$	$S_0(x_1)$	$S_1(x_1)$
0	0	2	3
1	0	3	2
2	0	0	1
3	0	1	0

y_1	$g_1(y_1)$
0	1
1	0
2	1
3	0

c)

Figure 8.
Illustration of the looping algorithm.

It is not difficult to generalize the above construction. As a first example, let us assume in figure 8 that $\Sigma_1 = \{0,1\}^2$. The network S in that figure may be replaced by the network in figure 9a, in which the g_i's are now elementary functions of u_1. If we analyze the resulting network by viewing it as a substitution network on

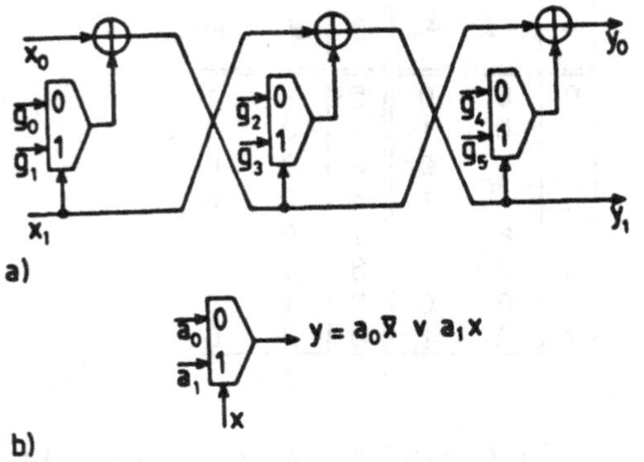

Figure 9.
Total substitution network on $\{0, 1\}^2$.

$\Sigma_2 \times \Sigma_1 \times \Sigma_0 = \{0,1\}^3$, we see that an arbitrary substitution on the considered alphabet may be realized as the cascade of 5 substitutions acting on a single coordinate as a function of the two other coordinates. The order of action on the coordinates is x_0, x_1, x_2, x_1, x_0, and the total substitution network on $\{0,1\}^3$ may thus be redrawn as shown in figure 10.

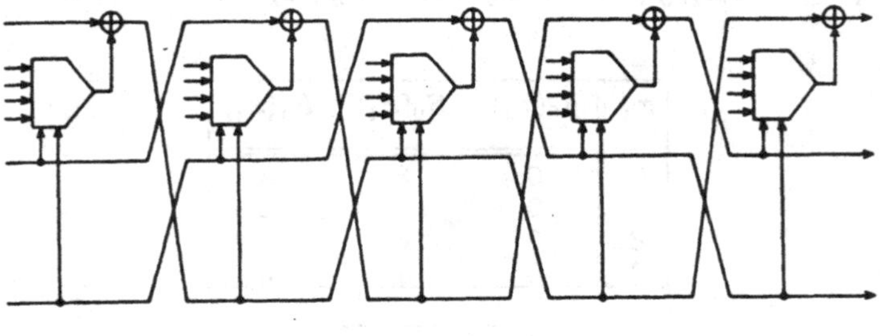

Figure 10.
Total substitution network on $\{0, 1\}^3$.

The expected generalization is now at hand: a total substitution network on $\{0,1\}^n$ may be built from $(2n-1)$ stages, each of which acts on a single coordinate by EX-ORing it with a function of the $(n-1)$ other coordinates. The order of the coordinates undergoing these successive changes is $x_0, x_1, \ldots, x_{n-1}, x_{n-2}, \ldots, x_0$. The number Γ of external inputs to this network is:

$$\Gamma = (2n-1)2^{n-1} = O(n2^n) = O(\log(2^n!)), \tag{14}$$

and it is asymptotically optimal.

Remark. It is interesting to note, from a bibliographical point of view, that the basic idea underlying the network in figure 10 is essentially the same as that used by

Good[18] in his study of fast transforms.

3.6. Conclusion

In the above paper, we established a link between various ways of viewing substitution networks: the shift register approach, the permutation network approach and the cryptographic approach. This obviously has the advantage of placing the study of substitution networks in a rather general frame. From a practical point of view, however, the above study is far from being complete. One should indeed:

- study easily invertible substitution networks, for example *involutive networks;*
- study the generating power of well chosen generating elements, such as the cell displayed in figure 5 (one-sided shift) and the interest of the generated permutation subgroups for cryptographical purposes;
- conversely, study the properties that should be owned by a substitution network to be secure in its cryptographical application.

References

1. A. G. Konheim, *Cryptography: A Primer*, J Wiley, New York (1981).
2. J. B. Kam and G. I. Davida, "Structured design of substitution-permutation encryption networks," *IEEE Transactions on computers* **C-28** pp. 747-753 (1979).
3. C. Ronse, "Substitution networks," *Philips Research Laboratory. Brussels* **R-444** (1980).
4. V. E. Benes, *Mathematical theory of switching networks and telephone traffic*, Academic press, New York (1965).
5. D. Slepian, "Two theorems on a particular switching network," *Unpublished manuscript*, (1952).
6. A. Waksman, "A permutation network," *Jl ACM* **15** pp. 159-163 (1968).
7. S. W. Golomb, *Shift register sequences*, Holden Day, San Francisco (1967).
8. C. E. Shannon, "Communication theory of secrecy systems," *BSTJ* **28** pp. 656-715 (1949).
9. H. Feistel, "Cryptography and computer privacy," *Scientific American*, pp. 15-23 (1973).
10. R. Morris, N. J. A. Sloane, and A. D. Wyner, "Assessment of the NBS proposed Data Encryption Standard," *Cryptologia* **1** pp. 301-306 (1977).
11. A. M. Whitehead, "Memoir on the algebra of symbolic logic," *Amer. Jl.of Math* **23** pp. 139-165; 297-316 (1901).
12. L. Lowenheim, "Gebietdeterminanten," *Math. Ann* **79** pp. 222-236 (1919).
13. S. Rudeanu, *Boolean functions and equations*, North Holland, Amsterdam (1974).
14. D. A. Huffman, "Canonical forms for information lossless finite state logical machines," *IRE Transactions on circuit theory* **CT-6** pp. 41-59 (1959). Special supplement

15. A. M. Duguid, "Structural properties of switching networks," *Brown University Progress report*, (1959).

16. V. J. Neiman, "Structure et commande optimales des reseaux de connexion sans bloquage," *Annales des telecommunications* **24** pp. 232-238 (1969).

17. N. T. Tsao-Wu, "On Neiman's algorithm for the control of rearrangeable switching networks," *IEEE transactions on communications* **COM-22** pp. 737-742 (1974).

18. I. J. Good, "The relationship between two Fast Fourier Transforms," *IEEE Transactions on computers* **C-20** pp. 310-317 (1971).

4. BASIC PROTOCOLS IN MODERN CRYPTOLOGY

In cooperation with J.-J. Quisquater
(Philips Research Laboratory Brussels, Belgium)

4.1. Introduction.

Classical cryptology has been mainly concerned with the *privacy* problem: its main goal is indeed to prevent an opponent from accessing confidential information. To reach this goal, it essentially use the ideas crystallized in the Shannon [1949] paper: basically, to resist cryptanalytical attacks based on language redundancy, a good cipher system should put at work both *transpositions* of letters within a message and alphabet *substitution*. Theses ideas are actually embedded in the DES (Data Encryption Standard) today widely recommended for protecting confidential commercial information. Observe that, in the classical approach, except in the DES context, the problems of key distribution, of identification and of signature are not discussed.

The background completely changes when business applications and specially banking applications start opening themselves to cryptological techniques. For example, if two mutually distrusting transaction partners wish to encipher one of their dialogues, they are likely to sep up a key building system in which they both play an essential role: this obviously restates the *key distribution* problem in completely new terms. Similarly, in a banking transaction, a message originator should be prevented from denying a posteriori having sent the message, and this raises the *signature* problem. As a matter of fact, modern cryptology adds to the privacy problem a new brand of challenges: the goal is now to prevent false information to be injected into a system. This is the *authentication* problem. The papers by Diffie & Hellman [1976, 1979], Merkle & Hellman [1978], Rivest, Shamir & Adleman [1978] provide the starting points of the new discussions.

The purpose of the present note is to review the main methods that have been proposed to date for the solution of these problems. In fact, these methods put at work two types of ingredients:

(i) *functions* endowed with specific properties (one-way functions, trapdoor one-way functions, ...) to be defined in the next section

(ii) sets of operating rules, which include computations with the preceding functions and transmission of messages over unsecure channels: these are the *protocols*.

4.2. Cryptographic functions. Notations and definitions.

In the present section, we first review the notational system introduced by Shamir [1980]. If we denote by D a data set, by R a result set and by S a user set, then:

$$D \xrightarrow[S]{} R \tag{1}$$

reads: the result R is easily deduced from the data D by all the members of the set S. Similarly, the notation

$$D \xrightarrow[S]{} R \tag{2}$$

is used to mean that, for every member of the set S, it is computationally unfeasible to retrieve the data D from the result R.

Let us denote by b the sender or message originator, by c the receiver and by g the opponent. Let us furthermore denote by $f_y(x)$ an encipherment function with cleartext x and key y. Then, the above notation may be illustrated by the following examples:

(i) $x, y \xrightarrow[b,\,c,\,g]{} f_y(x)$;
anyone who knows the key can encipher.

(ii) $f_y(x) \xrightarrow[b,\,c,\,g]{} x$;
the cleartext may not be extracted from the ciphertext.

...

Following Shamir (*loc. cit.*), we make the following two assumptions:

Assumption 1. The cleartexts, the ciphertexts and the keys are atomic objects taken from a common universe U.

Assumption 2. For each key $y \in U$, the encipherment function f_y is a permutation on U and thus has an inverse decipherment function f_y^{-1} satisfying:

$$\forall\, x, y \in U : f_y^{-1}(f_y(x)) = f_y(f_y^{-1}(x)) = x. \tag{3}$$

We now introduce our three main definitions.

Definition 1. The function f_y is a *one-way function* if and only if it satisfies the following axioms:

A1. $x, y \xrightarrow[b, c, g]{} z = f_y(x)$

A2. $z, y \xrightarrow[b, c, g]{} x = f_y^{-1}(z).$

For example, the function f_y could be the exponentiation over a finite field:

$$f_y : x \rightarrow y^x$$

with y a primitive field element. The function f_y could also be the raising to a power in a ring of integers

$$f_y : x \rightarrow x^y \pmod{n}, \quad n \text{ nonprime}.$$

There are many examples of one-way functions. It is indeed easy to understand that any cipher which is not broken by a known-text attack is a one-way function of its key. This is for example the case of the DES. If

$$z_i = DES_k(m_i),$$

then

$$(z_1, m_1), ..., (z_i, m_i), ... \xrightarrow[g]{} k.$$

Another important remark should be made about one-way function used in practice: when we say for example that the exponential $z = y^x$ is a one-way function of x, we mean that the best available algorithm for computing the logarithm $x = \log_y z$ over a finite field is significantly more complex than computing the exponential. This may be restated in terms of computational complexity: in the present state of the art, the computation of the logarithm has a much higher *upper* complexity *bound* than the computation of the exponential. The absolute security in cryptographic applications would obviously require a similar statement to hold true for *lower* complexity *bounds*.

Definition 2. The function f_y is a *trapdoor one-way function* if and only if it satisfies the following axioms:

B1. $x, y \xrightarrow[b, c, g]{} z = f_y(x)$

B2. $z, y \xrightarrow[c]{} x = f_y^{-1}(z)$

B3. $$z, y \xrightarrow[b,g]{} x = f_y^{-1}(z).$$

Thus, c detains a personal secret information which provides him with an efficient way of computing f^{-1} while no one else is able to deduce such an algorithm from the knowledge of the publicly revealed algorithm.

Three trapdoor one-way functions have been studied to date:

(i) the Rivest, Shamir, Adleman [1978] scheme, where the function f_y is the raising to a power in a finite ring modulo n and where the secret information ultimately rests upon the knowledge of the factorization of n.

(ii) the knapsack function (Merckle & Hellman [1978]).

(iii) the Mc Eliece [1978] scheme.

Let us briefly describe, for illustration purposes, the Rivest, Shamir, Adleman (RSA) technique. The preliminary operations are carried out by the privileged receiver, called c in the above notation; c performs the following computations:

(i) selection of two large primes p, q

(ii) computation of

$$n = p \times q$$

$$\phi(n) = (p-1)(q-1), \text{the Euler function of } n$$

(iii) choice of an integer e relatively prime with $\phi(n)$

(iv) computation of the multiplicative inverse d of e (mod $\phi(n)$). This number exists as $\gcd(e, \phi(n)) = 1$.

The numbers n, e are published and the encipherment is performed as:

$$x \mapsto z = x^e \pmod{n}.$$

The numbers d, p, q are kept secret by the receiver, who is able to perform the decipherment by

$$z \mapsto z^d = x^{ed} = x \pmod{n}.$$

The difficulty of decryption is due to the difficulty of factoring the large integer n: without knowing that factorization, an opponent has no simple method of extracting the eth root of z.

Clearly enough, trapdoor one-way functions may be used for encipherment and decipherment of confidential messages by entitled persons. From a practical point of view, however, the security of these systems imposes to the user numerical computations involving large numbers (500-bit numbers are currently recommended in the RSA scheme). This restricts the use of public-key encipherment to relatively low bit rates. These arguments provide some motivations to the research and study of additional classes of trapdoor one-way functions.

Definition 3. The function f_y is a *public key function* if and only if it satisfies the following axioms:

C1. $x, y \xrightarrow[b,c,g]{} z = f_y(x)$

C2. $z, x \xrightarrow[b,c,g]{} y$

C3. $\forall\, x, y, w : f_y(f_w(x)) = f_w(f_y(x))$.

Basically, axioms **C1** and **C2** express that f_y is a one-way function of the key. The usefulness of the commutative property **C3** will appear in the applications. The raising to a power is the only known public key function.

4.3. Key distribution.

Before starting an exchange of confidential information, two partners agree on a general encipherment-decipherment system. This system may be considered as publicly known and will have a relatively long lifetime. The two partners have then to share a secret and easily changed parameter, the key (or a system of related keys).

In its strongest form, the problem of key distribution may be stated as follows:

(i) at they outset, the two partners have no common information, except public information.

(ii) their only method of sharing information is to transmit messages over an unsecure communication channel.

A basic solution to this problem has been proposed by Diffie and Hellman [1976]. It rests upon the use of a public key function f_y and of a public parameter s. The key distribution protocol is illustrated by figure 1.

Observe that, as f_y is one-way with respect to the key, the transmission of $f_{k_b}(s)$ and of $f_{k_c}(s)$ does not reveal the secret parameters k_b and k_c.

Assume for example that the public key function is the exponential. Then, the numbers transmitted over the channel are s^{k_b} and s^{k_c} while the common key is $s^{k_b k_c} = s^{k_c k_b}$.

PARTNER b	TRANSMISSIONS	PARTNER c
SELECTS k_b		SELECTS k_c
COMPUTES $f_{k_b}(s)$		COMPUTES $f_{k_c}(s)$
	$\xrightarrow{f_{k_b}(s)}$	
	$\xleftarrow{f_{k_c}(s)}$	
COMPUTES		COMPUTES
$k = f_{k_b}(f_{k_c}(s))$		$k = f_{k_c}(f_{k_b}(s))$

Figure 1. *Key distribution protocol 1.*

Remarks.

(i) **The key distribution scheme in figure 1 is simplified by the existence of a public-key directory, i.e. by a set**

$$\{(b, f_{k_b}(s)), (c, f_{k_c}(s)), \ldots\}$$

in which each member of the system reveals his public parameter $f_{k_b}(s)$, $f_{k_c}(s)$, ... These quantities are thus no more transmitted over the channel, but are merely retrieved from the public file. That system is probably less probably less advisable than the preceding one as it only allows rather unfrequent key changes.

(ii) **The above method easily extends to the situation where a common key is required by more than two users. This problem, known as conference key distribution has been discussed by Ingemarsson [1980].**

PARTNER b	TRANSMISSIONS	PARTNER c
SELECTS k_b		SELECTS k_c
RETRIEVES f_c		RETRIEVES f_b
COMPUTES $f_c(k_b)$		COMPUTES $f_b(k_c)$
	$f_c(k_b)$ $\longrightarrow$	
	$f_b(k_c)$ $\longleftarrow$	
COMPUTES		COMPUTES
$k_c = f_b^{-1}(f_b(k_c))$		$k_b = f_c^{-1}(f_c(k_b))$
COMPUTES		COMPUTES
$k = g(k_b, k_c)$		$k = g(k_b, k_c)$

Figure 2. *Key distribution protocol 2.*

A second solution of the key distribution problem is based on trapdoor one-way functions. We assume here that each member b, c, ... of the system has selected a pair of functions (f_b, f_b^{-1}), (f_c, f_c^{-1}), ... and that a public directory

$$\{(b, f_b), (c, f_c), ...\}$$

has been issued. The protocol is then described in figure 2. It should be clear that b (resp. c) is the only person able to retrieve k_c (resp. k_b) from $f_b(k_c)$ (resp. $f_c(k_b)$). Hence, after the transmission, the partners b and c are the only persons to share the information $\{k_b, k_c\}$ from which the session key k is built as $k = g(k_b, k_c)$.

As in the former method, it is obviously possible to set up a variant of protocol 2 using no directory. More interesting is the fact that it is easy, using one-way trapdoor functions, to set up a dissymmetric system in which a single of the transaction partners chooses the session key. The process is illustrated in figure 3 which does not deserve further comments.

PARTNER a	TRANSMISSIONS	PARTNER b
SELECTS k		
RETRIEVES f_b		
COMPUTES $f_b(k)$		
	$\xrightarrow{f_b(k)}$	
		COMPUTES
		$k = f_b^{-1}(f_b(k))$

Figure 3. *Dissymmetric key distribution.*

Assume as an example that the trapdoor function in use is the RSA scheme. Thus b has published a public key (e_b, n_b) and obtains the secret parameters (d_b, p_b, q_b). Having selected k, the partner a would compute and transmit:

$$l = k^{e_b} \pmod{n_b}.$$

Partner b would then retrieve the key k as

$$k = l^{d_b} \pmod{n_b}.$$

An extensive discussion of the key distribution problem may be found in Merckle [1980]. We shall only observe here that this problem is greatly simplified if, at the outset of the transaction, the two partners detain some linked secret information. The situation may arise if, for example, the key distribution problem is linked with an authentication problem.

4.4. Identification.

The purpose of identification may be stated as follows: b wishes to convince himself that his transaction partner is actually c. This goal will be reached by:

(i) a sequence of transmissions over an unsecure channel

(ii) reference to some certified prepublished information. The basic idea to be put at work here is known as the *challenge* and *response* method: b will ask his transaction partner a question that c and c only can answer.

Let us also observe, before describing solutions to the identification problem, that an identification protocol should always be randomized: indeed, in the opposite situation, an opponent might impersonate c after having listened and recorded a previous identification dialogue.

PARTNER b	TRANSMISSIONS	PARTNER c
SELECTS r, a random		
RETRIEVES $(s_c, f_{k_c}(s_c))$		
COMPUTES $f_r(s_c)$		
	$f_r(s_c)$ $\longrightarrow$	
COMPUTES		COMPUTES
$\alpha = f_r(f_{k_c}(s_c))$		$\beta = f_{k_c}(f_r(s_c))$
	$f_{k_c}(f_r(s_c))$ $\longleftarrow$	
COMPARES α, β		

Figure 4. *Identification protocol 1.*

A first solution to the identification problem is grounded on a public key function f_y. Each of the system members, b, c, ... selects his own secret parameter k_b, k_c, ... and a directory $\{(b, (s_b, f_{k_b}(s_b))), (c, (s_c, f_{k_c}(s_c))), ...\}$ is published.

PARTNER b	TRANSMISSIONS	PARTNER c
SELECTS r_b, a random		SELECTS r_c, a random
RETRIEVES $(s_c, s_c^{k_c})$		RETRIEVES $(s_b, s_b^{k_b})$
COMPUTES $s_c^{r_b}$		
	$s_c^{r_b}$ $\longrightarrow$	
	$s_b^{r_c}$ $\longleftarrow$	
COMPUTES		COMPUTES
$\alpha = (s_c^{k_c})^{r_b}$		$\alpha' = (s_c^{r_b})^{k_c}$
$\beta = (s_b^{r_c})^{k_b}$		$\beta' = (s_b^{k_b})^{r_c}$
	α' $\longleftarrow$	
	β $\longrightarrow$	
COMPARES α, α'		COMPARES β, β'

Figure 5. *Symmetric identification protocol 1.*

The identification protocol is then described by figure 4. Let us briefly analyze this protocol. Observe first that, as f_y is one-way with respect to y, an opponent may not deduce r from $f_r(s_c)$. No one else than a can compute α. Partner c only, thanks to his knowledge of k_c, is able to compute β. Observe also that for the same reason, the secret parameter k_c may not be deduced from β.

Clearly enough, the protocol in figure 4 may be completed to yield a symmetric identification procedure by which each of the partners would identify his correspondent.

In this case, each of the two correspondents would generate a random number and challenge his partner. A symmetric identification protocol using the exponential is shown in figure 5. The public directory contains in this case the information $\{(b, (s_b, s_b^{k_b})), (c, (s_c, s_c^{k_c})), ...\}$.

It is also possible to ground the identification process on the use of trapdoor functions. We assume here that a public directory

$$\{(b, f_b), (c, f_c), ...\}$$

has been published. Each of the partners keeps secret his inverse function f_b^{-1}, f_c^{-1}, ... The identification of c by b then runs as shown in figure 6 which may obviously be completed to yield a symmetric identification procedure.

PARTNER b	TRANSMISSIONS	PARTNER c
SELECTS r, a random		
RETRIEVES f_c		
COMPUTES $f_c(r)$		
	$f_c(r)$ $\longrightarrow$	
		COMPUTES $r = f_c^{-1}(f_c(r))$
	r $\longleftarrow$	

Figure 6. *Identification protocol 2.*

More interesting to study is the case of the one-sided identification. We mean here that an authority a has to identify one of its subordinates, say b. This is typically an *access control* situation (see also Davio & Quisquater [1982]). The above procedures apply in this case, but it is noteworthy that a solution of the one-sided identification problem may be based on one way functions. That solution may be described as follows: the authority selects, and keeps secret:

(i) a one-way function f

(ii) a parameter s.

He furthermore selects a second one-way function, g. He then distributes to his subordinates:

(i) the function g

(ii) a signature $(k_b, f_{k_b}(s))$, $(k_c, f_{k_c}(s))$, ... of the function f.

The identification process is then described by figure 7.

AUTHORITY a	TRANSMISSIONS	PARTNER b
	k_b $\longleftarrow$	
COMPUTES $f_{k_b}(s)$		
SELECTS r, a random		
	r $\longrightarrow$	
COMPUTES		COMPUTES
$\beta = g_r(f_{k_b}(s))$		$\beta' = g_r(f_{k_b}(s))$
	β' $\longleftarrow$	
COMPARES β, β'		

Figure 7. *Access control.*

Observe

(i) that passive wiretapping does not compromise b's personal secret information $f_{k_b}(s)$, as g is one-way

(ii) that physical access to the information owned by b

a) does not compromise the system secret parameter s, as f is one-way

b) does not allow an opponent to impersonate a system member by creating a pair $(t, f_t(s))$ as neither f nor s are available to b

(iii) that it is possible, thanks to the use of two distinct one-way functions f and g, to differentiate the computational efforts required from a and b.

Conclusions.

The above discussion has introduced two of the problems raised by the modern applications of cryptology, *viz.* the key distribution problem and the identification problem. It furthermore shows the interest of special functions in the solution of these problems.

That discussion is however far from being closed: each practical situation offers its own specificities, suggests specific threats and attacks that should be handled appropriately. Furthermore, a number of important problems, such as the signature problem have not been considered here.

REFERENCES

Davio, M. and Quisquater, J. J., Methodology in information security. Mutual authentication procedures. Application to access control., *Proc. 1982 Zurich International Seminar on Digital Communication*, 1982, pp. 87-92.

Diffie, W. and Hellman, M. E., New directions in cryptography, *IEEE Trans. Inform. Theory*, **IT-22**, 6, Nov. 1976, pp. 644-654.

Diffie, W. and Hellman, M. E., Privacy and authentication. An introduction to cryptography, *Proc. IEEE*, **67**, 3, 1979, pp. 397-427.

Evans, A., Kantorovitz, W. and Weiss, E., A user authentication scheme not requiring secrecy in the computer, *Comm. ACM*, **17**, 1974, pp. 437-442.

Ingemarson, I., Tang, D. T. and Wong, C. K., A conference key distribution system, *IBM Research Report RC 8236* (#35599), 1980.

Ingemarson, I. and Wong, C. K., A user authentication scheme based on a trapdoor one-way function, *IBM Research Report*, 1980.

Mc Eliece, R. J., A public key cryptosystem based on algebraic theory, *Deep space network progress rept 42-44*, Pasadena, Jet propulsion lab., 1978, pp. 114-116.

Merkle, R. C., Protocols for public key cryptosystems, *Proc. 1980 conference on security and privacy*, IEEE, NY, 1980, pp. 122-134.

Merkle, R. C. and Hellman, M. E., Hiding information and signatures in trapdoor knapsacks, *IEEE Trans. Inform. Theory*, **IT-24**, 1978, pp. 525-530.

Rivest, R. L., Shamir, A. and Adleman, L., A method of obtaining digital signatures and public-key cryptosystems, *Comm. ACM*, **21**, Feb. 1978, pp. 120-126.

Shamir, A., On the power of commutativity in cryptography, in *"Automata, languages and programming", ICALP 80*, Lectures Notes in Computer Science n° 85, Springer-Verlag, Berlin, 1980, pp. 582-595.

Simmons, G. J., A system for point of sale or access user authentication and identification, *IEEE workshop on communication security*, Santa Barbara, CA., 1981.

5. KNAPSACK TRAPDOOR FUNCTIONS: AN INTRODUCTION

5.1. Introduction

The interest of trapdoor one-way functions in cryptology is well established. In the present introductory text, we study a specific function of this type, the trapdoor knapsack introduced in cryptology by Merkle and Hellman[1] After having recalled the definition of the knapsack problem and reviewed their most important properties, we discuss the main techniques available for the construction of trapdoor knapsacks and describe some of the arguments presented to date against the Merkle-Hellman trapdoor knapsacks.

5.1.1. Definitions.

A *knapsack system* is defined by

(i) an integer valued vector

$$\mathbf{a} = [a_{n-1},\ldots,a_1,a_0] : a_i \in \mathbf{N},\ \forall i \in \mathbf{Z}_n ;^{*}$$

(ii) a mapping

$$S_{\mathbf{a}} : \mathbf{Z}_2^n \to \mathbf{Z}_M : \mathbf{x} \to \sum_{i=0}^{n-1} x_i a_i ,$$

where $\mathbf{x}$ is the binary vector :

$$\mathbf{x} = [x_{n-1},\ldots,x_1,x_0] : x_i \in \mathbf{Z}_2 ,\ \forall i \in \mathbf{Z}_n ,$$

and

$$M = 1 + \sum_{i=0}^{n-1} a_i .$$

In general, a *knapsack problem* is stated as follows : given a knapsack system $<\mathbf{a}, S_{\mathbf{a}}>$ and an integer S, find, if any, the binary vector(s) $\mathbf{x}$ such that :

$$S = S_{\mathbf{a}}(\mathbf{x}) = \sum_{i=0}^{n-1} a_i x_i . \tag{1}$$

*. The symbol $\mathbf{Z}_a$ denotes the ring of integers modulo a.

A knapsack system $<\mathbf{a},S_\mathbf{a}>$ is *injective* if the mapping $S_\mathbf{a}$ is an injection. As usual in cryptography, we shall restrict ourselves in the present text to injective knapsacks. In the cryptological context, $\mathbf{x}$ denotes the cleartext, $S_\mathbf{a}(\mathbf{x})$ denotes the ciphertext and the knapsack problem may be restated as follows : given a knapsack system $<\mathbf{a},S_\mathbf{a}>$ and a ciphertext S, find the unique integer $\mathbf{x}$ satisfying (1). The knapsack problem is a classical mathematical problem (see e.g. Horowitz & Shani [2] and its application to cryptography was proposed by Merkle and Hellman[1] In general, the knapsack problem is known to be np-complete and this suggests the possibility of constructing trapdoor knapsacks.

5.2. Characterization of injective knapsacks.

Theorem 1. *A knapsack system* $<\mathbf{a},S_\mathbf{a}>$ *is injective iff :*

$$\forall I,J \subseteq \mathbf{Z}_n : I \cap J = \phi : \sum_{i\in I} a_i \neq \sum_{j\in J} a_j . \tag{2}$$

Equivalently, the knapsack system $<\mathbf{a},S_\mathbf{a}>$ *is injective iff, for any three-valued vector :*

$$\mathbf{c} = [c_{n-1},\dots,c_1,c_0] : c_i \in \{-1,0,1\},\ \forall i \in \mathbf{Z}_n ,$$

the inequality

$$\sum_{i=0}^{n-1} c_i a_i \neq 0 \tag{3}$$

holds.

Proof. We first discuss condition (2).

(i) *Necessity.* If condition (2) is violated, it is obviously possible to select two binary n-tuples $\mathbf{x}^{(1)}$ and $\mathbf{x}^{(2)}$ such that $S_\mathbf{a}(\mathbf{x}^{(1)}) = S_\mathbf{a}(\mathbf{x}^{(2)})$.

(ii) *Sufficiency.* Assume that condition (2) is satisfied. We select two distinct binary vectors $\mathbf{x}^{(1)}$ and $\mathbf{x}^{(2)}$ and show that $S_\mathbf{a}(\mathbf{x}^{(1)}) \neq S_\mathbf{a}(\mathbf{x}^{(2)})$. Let $I = \mathbf{x}^{(1)} \cap \overline{\mathbf{x}^{(2)}}$ and $J = \overline{\mathbf{x}^{(1)}} \cap \mathbf{x}^{(2)}$ (Boolean operations apply componentwise). As condition (2) applies for these particular I and J, we have obviously $S_\mathbf{a}(\mathbf{x}^{(1)}) \neq S_\mathbf{a}(\mathbf{x}^{(2)})$.

Let us now show that conditions (2) and (3) are equivalent :

(i) assume first that, for some $I,J \subseteq \mathbf{Z}_n$ such that $I \cap J = \phi$ one has:

$$\sum_{i\in I} a_i = \sum_{j\in J} a_j . \tag{4}$$

This is clearly equivalent to

$$\sum_{i=0}^{n-1} c_i a_i = 0, \tag{5}$$

if one defines the vector $\mathbf{c}$ by

$$\begin{cases} c_k = 1 & \textit{iff } k \in I, \\ c_k = -1 & \textit{iff } k \in J, \\ c_k = 0 & \textit{otherwise}; \end{cases}$$

(ii) assume now that equation (5) holds. Define disjoint subsets $I, J \subseteq \mathbf{Z}_n$ by

$$I = \{k \mid c_k = 1\},$$
$$J = \{k \mid c_k = -1\}.$$

For these sets, the equality (4) is obviously satisfied.

■

In the present state of our knowledge, it seems difficult to check the injective nature of a given knapsack by non-enumerative techniques : brute force approachs would indeed enumerate the 2^n vectors $\mathbf{x}$ or the 3^n vectors $\mathbf{c}$. These methods are obviously of exponential complexity.

5.3. Easy knapsacks

In some cases however, the injective nature of a knapsack system is easily detected and the corresponding knapsack problem is at the same time easy to solve. The knowledge of these knapsacks is fundamental as the possibility of trapdoor knapsacks is actually based on the existence of *easy* and *difficult* knapsacks and on transformations relating these two types of knapsacks.

As an intermediate concept, we shall use in what follows the notion of reducible knapsack : a knapsack $<\mathbf{a}, S_\mathbf{a}>$ is reducible if, given $S = S_\mathbf{a}(\mathbf{x})$, it is possible to determine one of the components x_i of $\mathbf{x}$.

5.3.1. Superincreasing knapsacks

A first type of reducible knapsack may be described as follows :

$$\text{for all } i \in \mathbf{Z}_n : a_i > \sum_{j \neq i} a_j. \tag{6}$$

Let indeed :

$$M = \sum_{j \neq i} a_j.$$

One sees that if $S > M$, then $x_i = 1$, else $x_i = 0$.

The knapsack $<\mathbf{a}, S_\mathbf{a}>$ is *superincreasing* iff:

(i) it satisfies condition (6)

(ii) the residual knapsack $\mathbf{b} = \mathbf{a} \setminus \{a_i\}$, obtained by deleting from $\mathbf{a}$ the component a_i is itself superincreasing.

The knapsack $<\mathbf{a}, S_\mathbf{a}>$ is *superincreasing in the strict sense* iff :

$$\forall i \in \mathbf{Z}_n : a_i > \sum_{j<i} a_j. \tag{7}$$

Stated otherwise, if $<\mathbf{a}, S_\mathbf{a}>$ is superincreasing in the strict sense, it is superincreasing and its components are ordered according to $a_0 < a_1 < \cdots < a_{n-1}$.

As an example, define the knapsack $\mathbf{a}$ by

$$a_i = 2^i, \ \forall i \in \mathbf{Z}_n. \tag{8}$$

The corresponding knapsack is superincreasing and $\mathbf{x}$ is the (unique) binary representation of $S_a(\mathbf{x})$.

We now state without proof the following property :

Theorem 2. *A Superincreasing knapsack is injective.*

The proof of theorem 2 is elementary and ultimately rests upon the observation that any subset of a superincreasing knapsack is itself superincreasing.

As a matter of fact, it is quite easy to generate superincreasing knapsacks, to test a given knapsack for superincreasingness and to solve the knapsack problem in the superincreasing case. An elementary complexity study would show that all these operations may be carried out in polynomial time.

Remark. Let $<\mathbf{a}, S_\mathbf{a}>$ be a strictly superincreasing knapsack and consider some a_i. Then:

$$a_{i+1} \geq a_i + 1$$

$$a_{i+2} \geq a_{i+1} + a_i + 1 \geq 2a_i + 2$$

$$a_{i+3} \geq a_{i+2} + a_{i+1} + a_i + 1 \geq 4a_i + 4$$

....

$$a_{i+j} \geq 2^{j-1}a_i + 2^{j-1}.$$

This means that, from a_{i+2} on, the number of bits of the binary representation of a_{i+j} increases by one unit (at least) at each unit increase of j. A typical bit profile corresponding to that situation is shown in figure 1.

a_0	5	101
a_1	6	110
a_2	12	1100
a_3	24	11000
a_4	65	1000001
a_5	117	1110101
a_6	230	11100110
a_7	460	111001100
a_8	923	1110011011

Figure 1.
Bit profile in a superincreasing knapsack

5.3.2. Divisibility knapsacks

Superincreasing knapsacks are by far the most popular easy knapsacks. They are however not the only easy knapsacks, as we shall illustrate in the present section. We first describe a second type of reducible knapsack. Assume indeed that the knapsack $<\mathbf{a}, S_\mathbf{a}>$ satisfies the following condition :

$$\forall i \in \mathbf{Z}_n; \quad d \in \mathbf{N} : d \mid a_i \text{ and } d \nmid a_j, \; \forall j \neq i. \tag{9}$$

This knapsack is clearly reducible. Indeed, if $d \mid S$, then $x_i = 0$, else $x_i = 1$.

Now, the knapsack $<\mathbf{a}, S_\mathbf{a}>$ is a *divisibility knapsack* iff it satisfies condition (9) for all $i \in \mathbf{Z}_n$. To construct divisibility knapsacks, one may use a method similar to that used in the Chinese remainder theorem. Consider indeed a series $q_{n-1},\ldots,q_1,q_0$ of pairwise prime numbers. Compute :

$$Q = \prod_{i=0}^{n-1} q_i; \tag{10}$$

$$r_i = Q/q_i = q_{n-1}\cdots q_{i+1}q_{i-1}\cdots q_0. \tag{11}$$

The knapsack $<\mathbf{r},S_\mathbf{r}>$ is a divisibility knapsack. Indeed if

$$S_\mathbf{r}(\mathbf{x}) = \sum_{i=0}^{n-1} r_i x_i, \tag{12}$$

then :

$$S_\mathbf{r}(\mathbf{x}) \equiv r_k x_k \mod(q_k).$$

as all the r_i's are divisible by q_k except r_k.

Numerical example.

$$\mathbf{q} = [q_5,q_4,q_3,q_2,q_1,q_0] = [13,11,7,5,3,2].$$

$$Q=30030.$$

$$\mathbf{r} = [r_5,r_4,r_3,r_2,r_1,r_0] = [2030,2730,4290,6006,10010,15015].$$

Let $\mathbf{x} = [0,1,1,0,1,0]$. The encipherment, carried out by (12) yields $S_\mathbf{r}(\mathbf{x}) = 17030$. Conversely, given $S = 17030$, one may decipher by (13) :

$17030 = 0 \bmod 13 \rightarrow x_5 = 0.$

$17030 = 2 \bmod 11 \rightarrow x_4 = 1.$

$17030 = 6 \bmod 7 \rightarrow x_3 = 1.$

$17030 = 0 \bmod 5 \rightarrow x_2 = 0.$

$17030 = 2 \bmod 3 \rightarrow x_1 = 1.$

$17030 = 0 \bmod 2 \rightarrow x_0 = 0.$

Remark. In the preceding discussion, we described two types of reducibility yielding two types of easy knapsacks, viz. the superincreasing knapsacks and the divisiblility knapsacks. Clearly, the two types of reducibility could be combined to yield mixed easy knapsacks.

5.4. Similarity transformations

Consider an invertible transformation $T : \mathbf{N} \rightarrow \mathbf{N}$. Let $<\mathbf{a},S_\mathbf{a}>$ represent a knapsack. Its transform by T is the knapsack $<\mathbf{b},S_\mathbf{b}>$ with $b_i = T(a_i),\ \forall i \in \mathbf{Z}_n$. The transform T is a ***similarity transformation*** iff

$$S_\mathbf{b}(\mathbf{x}) = T(S_\mathbf{a}(\mathbf{x})). \tag{14}$$

The interest of the similarity transformations is obvious. Indeed, if one is able to solve the knapsack problem for either one of two similar knapsacks, one is immediately provided with a solution of the knapsack problem for the second one.

In practice, one most often uses similarity transformations to built apparently difficult knapsacks from easy ones. The similarity transformation is then kept secret.

It is interesting to observe that similarity transformations may also be used for cryptanalytical purposes. Indeed, given an apparently difficult knapsack **a**, one might search for a similarity transformation T mapping **a** on an easy knapsack **b**.

The Merkle - Hellman transformation. Consider the knapsack $<\mathbf{a}, S_{\mathbf{a}}>$ and select two integers w, m such that :

$$g.c.d.(w,m) = 1. \tag{15}$$

Define the transformation T by

$$T : a_i \to b_i = a_i w \bmod \mathrm{m}. \tag{16}$$

As w and m are relatively prime, the transformation T is invertible ; more precisely

$$T^{-1} : b_i \to a_i = b_i w^{-1} \bmod \mathrm{m}. \tag{17}$$

Furthermore,

Theorem 3. *The transformation T is a similarity transformation if*

$$m > \sum_{i=0}^{n-1} b_i.$$

Proof. *

$$S_{\mathbf{a}}(\mathbf{x}) = \sum_{i=0}^{n-1} a_i x_i.$$

$$\begin{aligned} T(S_{\mathbf{a}}(\mathbf{x})) &= [w \sum_{i=0}^{n-1} a_i x_i]_m \\ &= [\sum_{i=0}^{n-1} w a_i x_i]_m \\ &= [\sum_{i=0}^{n-1} [w a_i]_m x_i]_m \\ &= [\sum_{i=0}^{n-1} b_i x_i]_m \\ &= \sum_{i=0}^{n-1} b_i x_i. \end{aligned}$$

■

In the practical application, **b** would be an easy knapsack, say a superincreasing knapsack, and it would be kept secret, together with w and m. The a_i's only would be published. To the message **x** would then correspond the cryptogram $S_{\mathbf{a}}(\mathbf{x})$ and the decipherment would then compute:

$$S_{\mathbf{b}}(\mathbf{x}) = T(S_{\mathbf{a}}(\mathbf{x})) = S_{\mathbf{a}}(\mathbf{x}) w \bmod \mathrm{m}. \tag{18}$$

Numerical example.

* The notation $[a]_m$ represents the smallest nonnegative residue of a mod m.

$\mathbf{b} = [b_0,b_1,b_2,b_3,b_4,b_5] = [89,150,392,754,1541,3235]$;

b is superincreasing.

$w = 107$; $m = 6473$ (a prime); w^{-1} mod m = 121.

Hence:

$\mathbf{a} = [a_0,a_1,a_2,a_3,a_4,a_5] = [3050,3104,3106,3002,3062,3076]$.

Let $\mathbf{x} = [x_0,\ldots,x_5] = [0,1,0,1,1,0]$.

$S_\mathbf{a}(\mathbf{x}) = 3104 + 3002 + 3062 = 9168$.

$S_\mathbf{b}(\mathbf{x}) = 9168 \,.\, 121 \bmod 6473 = 2445$.

$2445 < 3225 \to x_5 = 0$;

$2445 > 1541 \to x_4 = 1$; $2445 - 1541 = 904$.

$914 > 754 \to x_3 = 1$; $904 - 754 = 150$.

$150 < 392 \to x_2 = 0$.

$150 \geq 150 \to x_1 = 1$; $150 - 150 = 0$.

$0 < 89 \to x_0 = 0$.

■

Remark.

It is possible to generalize the concept of similarity transformation. Consider indeed two transformations T and U on **N** satisfying the two following conditions

$$\forall i \in \mathbf{Z}_n,\ a_i = T(b_i) \tag{19}$$

$$S_\mathbf{b}(\mathbf{x}) = U(S_\mathbf{a}(\mathbf{x})). \tag{20}$$

In such a situation, it will again be possible, starting from an easy knapsack **b** to create an apparently difficult knapsack **a**. It will afterwards be possible to decipher $S_\mathbf{a}(\mathbf{x})$ by retrieving first $S_\mathbf{b}(\mathbf{x})$ and using then the easy nature of **b**.

As an example of the above method, let us select an integer m and an integer vector **r** (possibly with negative components) and let us define

$$T : b_i \to a_i = r_i m + b_i. \tag{21}$$

$$U : S_\mathbf{a}(\mathbf{a}) \to [S_\mathbf{a}(\mathbf{x})]_m = S_\mathbf{b}(\mathbf{x}). \tag{22}$$

It is now immediate to check that, if

$$m > \sum_{i=0}^{n-1} b_i \tag{23}$$

the transformations (21) and (22) satisfy (19) and (20). The proof is identical to that given above for the Merkle-Hellman transformation.

Numerical example.

$[b_0,b_1,b_2,b_3] = [5,12,31,53]$\{(superincreasing)\}

$m = 135 > b_0 + b_1 + b_2 + b_3 = 101$.

$[r_0, r_1, r_2, r_3] = [6,4,3,-2]$.

$[a_0, a_1, a_2, a_3] = [815,552,436,-217]$.

Let $\mathbf{x} = [1,0,1,1]$. Then

$S_a(\mathbf{x}) = 815+436-217 = 1034$.

Thus

$S_b(\mathbf{x}) = [1034]_{135} = 89$.

Now

$89 \geq 53 \rightarrow x_3 = 1; \; 89 - 53 = 36$.

$36 < 31 \rightarrow x_2 = 1; \; 36 - 31 = 5$.

$5 < 12 \rightarrow x_1 = 0$.

$5 \geq 5 \rightarrow x_0 = 1$.

In practice, one will often choose $m = 2^t$ for some suitable exponent t. The residuation with respect to m then becomes a truncation of the binary representation. With the binary system in mind, Graham and Shamir, quoted in Shamir and Zippel[3] proposed a knapsack structure in which decoding becomes trivial. Let

$$a_i = \sum_{j=0}^{N-1} e_{i,j} 2^j . \tag{24}$$

Form the matrix $E = [s_{i,j}]$ by numbering the columns from right to left in such a way that row i of the matrix is the binary representation of a_i, read in the usual way. Assume now that the matrix E has the following type and dimensions :

$$E = \begin{bmatrix} \text{RANDOM} & & & \\ R_1 & 1_n & \begin{matrix} \text{BUFFER} \\ 0 \end{matrix} & \begin{matrix} \text{RANDOM} \\ R_2 \end{matrix} \\ \underbrace{}_{r_1} & \underbrace{}_{n} & \underbrace{}_{s} & \underbrace{}_{r_2} \end{bmatrix}$$

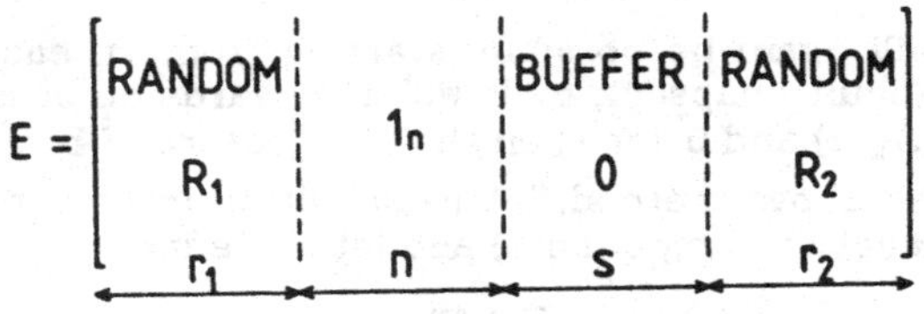

with $N = r_1 + n + s + r_2$. The submatrix 1_n is the unit matrix of order n and the width s of the buffer zone is chosen in such a way that the sum of the n quantities in the R_2 zone cannot overflow in the unit matrix. As this sum is upper bounded by $n 2^{r_2}$, its binary representation requires at most $r_2 + \lceil log_2 n \rceil$ * bits so that the choice $s = \lceil log_2 \rceil$ is convenient for the purpose.

Clearly now, if

$$S_a(\mathbf{x}) = [s_{N-1}, \ldots, s_{n+s+r_2} | s_{n+s+r_2-1}, \ldots, s_{s+r_2} | 0, \ldots, 0 | s_{r_2-1}, \ldots, s_0]$$

then

---
* The notation $\lceil a \rceil$ represents the smallest integer non smaller than a.

$$\mathbf{x} = \left[s_{n+s+r_g-1},\ldots,s_{s+r_g}\right].$$

In this case, the deciphering transformation U is reduced to a bit extraction.

5.5. Criticisms against the knapsack trapdoor functions.

As most of the recently proposed cryptographic functions, the knapsack trapdoor functions have been (and are) carefully studied. Two types of criticisms have resulted from these studies.

A. *Practical considerations.*

The practical use of trapdoor knapsacks is made difficult for two reasons : the size of the public key (the apparently difficult knapsack **a**) and the bit expansion : the latter term simply means that the cryptogram $S_{\mathbf{a}}(\mathbf{x})$ has more bits than the message **x** itself. To illustrate this point, let us recall the orders of magnitude originally suggested by Merkle and Hellman

$n = 100.$

$2^{101} < m < 2^{102}.$

$(2^i - 1) \,.\, 2^{100} < b_i < 2^i 2^{100}.$

$1 < w < m - 1.$

These choices insure that **b** is superincreasing, that the condition of theorem 3 is satisfied and that exhaustive search is impossible as each parameter has 2^{100} possible values. However, the key size is 20 kbits. Furthermore, each a_i will require 202 bits, and $S_{\mathbf{a}}(\mathbf{x}$ will thus require 209 bits, yielding an expansion of 2,09.

B. *Theoretical considerations.*

Basically, the criticism here is that trapdoor knapsacks are far from being theoretically certified as secure cryptographic functions. In fact, we ignore if a knapsack **a** obtained from an easy knapsack **b** by some transformation T^{-1} is really difficult to solve without the knowledge of the transformation T.

The point is illustrated by an example borrowed from a recent paper by Desmedt and al. [4] Consider the knapsack **a** = [5457,1663,216,6013,7439]. If we multiply this knapsack by 3950 mod 8443, we obtain the superincreasing knapsack **b** = [171,196,457,1191,2410]. However, after multiplying **a** by 46 mod 77, we obtain $\mathfrak{b}$=[2,37,3,14,6], which is also superincreasing (while not strictly). As **b** and $\mathfrak{b}$ both satisfy the condition of theorem 3, the two above transforms are similarity transforms and may thus be used for easy decoding. It is in fact shown that, if an apparently difficult knapsack **a** has been obtained from a superincreasing knapsack **b** by a Merkle-Hellman transformation, there are infinitely many ***trapdoor pairs*** (w,m) defining a similarity transform from **a** to superincreasing knapsacks **b**,$\mathfrak{b}$,... and the role of the cryptanalyst reduces to find one such pair.

Thus, if we know that **a** is obtained from a superincreasing **b** by a Merkle-Hellman transformation, we could think of the following method of attack : find W,M and compute $\mathbf{B} = W\mathbf{a} \bmod M$ such that :

$$(i)\ \forall i : B_i > \sum_{j \neq i} B_j$$

$$(ii)\ M > \sum_{i=0}^{n-1} B_i .$$

In that case, the knapsack **B** would be reducible and the process could be iterated to solve completely the **a** knapsack problem. In a recent research announcement, Shamir [5] claims to have solved this problem.

References

1. R. C. Merkle and M. E. Hellman, "Hiding information and signatures in trapdoor knapsacks," *IEEE transactions on information theory* **24** pp. 525-530 (1978).
2. E. Horowitz and S. Sahni, "Computing partitions with applications to the knapsack," *Jl of the ACM* **21** pp. 277-292 (1974).
3. A. Shamir and R. E. Zippel, "On the security of the Merkle-Hellman cryptographic scheme," *IEEE transactions on information theory* **IT-26** pp. 339-340 (1980).
4. Y. Desmedt, J. Vandewalle, and R. Govaerts, "Critical analysis of the Knapsack Public Key Algorithm," *IEEE Transactions on information theory*, (1982). to appear
5. A. Shamir, *A polynomial time algorithm for breaking Merkle-Hellman cryptosystems*, The Neiman Insititute, Rehovot, Israel (1982). Research announcement; preliminary draft

6. THE RSA TRAPDOOR FUNCTION: AN ANALYSIS OF SOME ATTACKS

In cooperation with C. Couvreur
(Philips Research Laboratory Brussels, Belgium)

6.1. Introduction

Since the publication by Rivest, Shamir and Adleman [1] of the basic principles of their public-key cryptosystem (usually referred to as the RSA or M.I.T. cryptosystem), a number of papers appeared in the literature, pointing out possible weaknesses, proposing some attacks based on these, and sometimes discussing possible countermeasures. In this section we briefly analyse the most important of these.

The attack based on possible decryption by factoring n is first discussed. Then, decryption by iteration, another possible attack, is analysed, and finally, we discuss the problem of choosing the key parameters so as to minimize the chance of success for these attacks, and we conclude with a few remarks.

For a description of the basic principles of the RSA trapdoor function, we refer to Section 4, Definition 2.

6.2. Decryption by factoring n

A cryptanalyst could try to find an integer x, not relatively prime to n. Then using Euclid's algorithm he could compute $gcd(x,n)$ and find either p or q, thus breaking the system. However, between 0 and $n-1$, there are $\Phi(n)$ numbers relatively prime to n and $n-\Phi(n) = p+q-1$ numbers not relatively prime to n. Thus the chance of finding an integer x having p or q as a factor is

$$\frac{n-\Phi(n)}{n} = \frac{p+q-1}{pq} = \frac{1}{p} + \frac{1}{q}$$

which is extremely small for large values of p and q.

We also observe that the choice $p=q$ is unacceptable, since p could then be obtained by taking the square root of the publicly known modulus n.

Moreover even if the difference $(p-q)$ is nonzero, $(p-q)$ must still be unpredictable, otherwise p and q would be determined from n by the difference of squares method of factoring.

A large number of factoring methods exist, but the fastest currently available, due to Morrison and Brillhart [2] and modified by Schroeppel, permits now to factor n in a number of steps approximately equal to

$$n \sqrt{\log(\log n)/\log n}.$$

For example, if n is about 10^{200}, $1.2 * 10^{23}$ operations are needed to factor n and if each operation uses one microsecond, the required time is $3.8 * 10^{9}$ years[1]

Another method, due to Pollard [3] factors n in $0(p^{1/2})$ operations, where p is the smallest prime factor of n. This explains why p and q should be large (each about 10^{100}) and about the same order of magnitude. (However $p-q$ should not be too small).

Mention should be made of other methods, briefly discussed in Williams and Schmid[4] which are frequently successful when a prime divisor p of n has the property that either $p+1$ or $p-1$ has only small prime factors. Thus, the prime factors p,q of n should be selected so that $p-1, p+1, q-1, q+1$ all have at least one large prime divisor. In particular the RSA cryptosystem will be more secure when $p = ap' + 1$ and $q = bq' + 1$ and p', q' are distinct large prime numbers.

Another method can also be suggested. Let x be a positive integer, which we may assume is relatively prime with n. Then its order modulo p or q (noted $ord_p x$ or $ord_q x$ respectively) may be a divisor of a or b. A possible attack would then consist in testing $gcd(n, x^i - 1)$ for $i = 1,2...m$, where m is a moderate-sized integer, with the hope that this quantity by greater than 1.

To resist this attack the designer of a RSA cryptosystem should choose a and b sufficiently small. For example, if $a = b = 2$, the probability that $gcd\ (n, x^2-1) \neq 1$ is about 10^{-100} (if p and q are each about 10^{100}).

Consequently more protection can be achieved if the following conditions are fulfilled:

(1) p and q are each about 10^{100} and differ in length by only a few bits.

(2) $p = ap' + 1$ and $q = bq' + 1$, where p', q' are large prime numbers and a, b are small.

(3) both $p + 1$ and $q + 1$ contain a large prime factor.

6.3. Decryption by iteration

The following cryptanalytic attacks are based on iteration of the encryption procedure. They allow decryption without factoring n.

In the attack devised by Simmons and Norris [5] one tries to find, for a given encrypted message $C = M^e \bmod n$, an integer j such that

$$C^{e^j} \bmod n = C \tag{2}$$

Then C can be decrypted, since

$$C^{e^{j-1}} \bmod n = M.$$

The following small example illustratres this attack:

Let $p = 983$, $n = 383*563 = 215629$, with $e = 49$, $d = 56956$.
Let the message be $M = 123456$.
The cryptanalyst knows the encrypted form of the message C

$$C = M^{49} \bmod n$$
$$= 1603.$$

He computes successively

$C_1 = C^{49} \bmod n = 180661$

$C_2 = C_1^{49} \bmod n = 109265$

$C_3 = C_2^{49} \bmod n = 131172$

$C_4 = C_3^{49} \bmod n = 98178$

$C_5 = C_4^{49} \bmod n = 56372$

$C_6 = C_5^{49} \bmod n = 63846$

$C_7 = C_6^{49} \bmod = 146799$

$C_8 = C_7^{49} \bmod n = 85978$

$C_9 = C_8^{49} \bmod n = 123456 = M$

$C_{10} = C_9^{49} \bmod n = 1603 = C.$

Of course j will be a factor of the order of e modulo $\varphi(n)$. In the example, j=10, which is actually equal to the order of e modulo $\varphi(n)$.

The attack of Simmons and Norris was generalized by Herlestam[6] (Note that both of these were criticized by Rivest)[1,7]
We briefly describe Herlestam's idea :
Let M be a message and $C = M^e \bmod n$ its encrypted form. If one can find a polynomial $P(e) = eQ(e)$ such that

$$C^{P(e)} \equiv C \bmod n \tag{3}$$

then C can be decrypted, since

$$M = C^{Q(e)} \bmod n.$$

6.4. Selection of the key parameters.

The key is determined by the choice of the prime factors, p and q, of the modulus n and the coding exponent e. In this section we briefly discuss how these parameters should be selected so as to minimize the chance of success of a cryptanalyst in his attempt to decrypt, either by factoring n or by an iterative method.

We have seen that to make the system resistant to decryption by factoring n, the prime factors p and q should be selected so that

(i) p and q are each about 10^{100} (but their difference should not be too small);

(ii) $p = ap' + 1$ and $q = bq' + 1$, where p', q' are both large prime numbers and a, b are small;

(iii) $p+1$ and $q+1$ each contain at least one large prime factor.

So, let us assume that p and q satisfy the above requirements. Then, as pointed out by Williams and Schmid [4] in order to break the cipher C by an iterative method, the cryptanalyst must fnd a polynomial $P(x)$ such that

$$P(e) \equiv 0 \bmod p'$$

or

$$P(e) \equiv 0 \bmod q'.$$

If $P(x)=F(x)=x-1$, as suggested by Simmons and Norris[5] it is necessary to find k such that

$$e^k \equiv 1 \bmod p'$$

or

$$e^k \equiv 1 \bmod q'.$$

To make this attack impractible one must ensure that k is quite large. One way of doing this, as suggested by Rivest [1] is to choose

$$p' = a'p'' + 1,\ q' = b'q'' + 1,$$

where p'', q'' are large prime numbers and a', b' small. In this case it is very likely that p'' (or q'') will divide k and hence k will be large. However, the actual process of finding $p = ap' + 1$, $q = bq$ tedious. Therefore, Williams amd Schmid suggested another approach, consisting in properly selecting the coding exponent e so as to make sure that the orders of e mod p' and mod q' are sufficiently large. The idea is this. Suppose $p' - 1 = fs$, $q' - 1 = ht$, where the complete factorizations of f and h are known, and s, t are composite but with no prime factor less than a bound b. If e is selected so as to satisfy:

$$e^{(p'-1)/s} \equiv 1 \bmod p',$$

and

$$e^{(p'-1)/r} \equiv 1 \bmod p',$$

for each prime factor r of f, then

$$ord_{p'}(e) > bf.$$

Similarly, if

$$e^{(q'-1)/t} \equiv 1 \bmod q',$$

and

$$e^{(q'-1)/r} \equiv 1 \bmod q',$$

for each prime factor r of h, then

$$ord_{q}'(e) > bh.$$

The designer can fix b quite large and since $ord_p'(e)$ and $ord_q'(e)$ divide k, k will be large too. Williams and Schmid further observe that it is auite possible to select the coding exponent e so as to satisfy the above requirements, and they showed that the number of such e is not very restricted. The reader is referred to their paper [4] for further details and some computational results concerning the construction of key sets.

Let us note in passing that it might be desirable to select e so as to satisfy $e > \log_2 n$, this in order to ensure that every message $M > 1$ is sufficiently scrambled by the reduction modulo n in the encryption process.

As a final remark, we quote the result* of Blakley and Borosh [8] that for every coding exponent e there exist at least nine messages M which are lft unchanged by the encryption process, that is, satisfying

$$M^e \equiv mod\ n.$$

This can be seen as follows. Since e is relatively prime with $\varphi(n)$ e is odd, and so each of the two congruences

$$x^e \equiv x \bmod p,$$
$$x^e \equiv x \bmod q,$$

has the three solutions x=0,1 and -1. Combining these by the Chinese Remainder Theorem, we obtain nine values for M. For example, for n = 47*59 = 2773, these are : 0, 1, 2772, 2537, 2303, 235, 2538, 471, and 236. Of course, publicly revealing any of the six nontrivial (i.e. different from 0, 1 or - 1) unconcealable messages would allow factoring n. It should be noted that there might be more than nine unconcealable messages. Their number N, as shown by Blakley and Borosh [8] is given by

$$N = (1+gcd(e-1)) * (1+gcd(e-1,q-1)).$$

thus n=9 if and only if

$$gcd(e-1,\lambda(n)) = 2.$$

6.5. Concluding remarks.

With regard to presently available methods of attack, the RSA public - key cryptosystem can be made secure by properly selecting the key parameters. However the actual process of generating appropriate primes p and q selecting the coding exponent might prove to be a rather tedious and lengthy task, especially if the scheme is to be used on a large scale. Moreover, it should be pointed out that the encryption and decryption operations (exponentiation modulo a large number) are rather complex, so that hardware implementation should probably be limited to relatively slow data rates (not exceeding a few kilobits per second).

Finally it should be noted that this system has not yet been *proved* to be secure.

* A similar result was obtained by D.R. Smith and J.T. Palmer (" Universal fixed messages and the Rivest-Shamir-Adleman cryptosystem ", Mathematika 26 (1979), 44-52).

References

1. R. L. Rivest, "Remarks on a proposed cryptanalytic attack on the MIT public-key cryptosystem," *Cryptologia*, pp. 62-65 (1978).
2. M. A. Morrison and J. Brillhart, "A method for factoring and the factorization of F7," *Math. Comp.* **29** pp. 183-205 (1975).
3. J. H. Pollard, "A Monte-Carlo Method for Factorization," *BIT* **15** pp. 331-334 (1975).
4. H. C. Williams and B. Schmid, "Some remarks concerning the MIT public-key cryptosystem," *BIT* **19** pp. 525-538 (1979).
5. G. J. Simmons and M. J. Norris, "Preliminary comments on the MIT public-key cryptosystem," *Cryptologia* **1** (4) pp. 406-414 (1977).
6. T. Herlestam, "Critical remarks on some public-key cryptosystems," *BIT* **18** pp. 493-496 (1978).
7. R. L. Rivest, by T. Herlestam"" "Critical remarks on "Critical Remarks on some public-key cryptosystems" by T. Herlestam," *BIT* **19** pp. 274-275 (1979).
8. G. R. Blakley and I. Borosh, "Rivest-Shamir-Adleman public-key cryptosystems do not always conceal messages," *Computers and Mathematics with Applications* **5** pp. 169-178] (1979).

References

- H. Beker and F. Piper, *Cipher systems*, Northwood Books, London (1982).
- V. E. Benes, *Mathematical theory of switching networks and telephone traffic*, Academic press, New York (1965).
- B. Blakley and G. R. Blakley, "Security of number theoretic public-key cryptosystems against random attack, II," *Cryptologia* **1** pp. 29-41 (1979).
- G. R. Blakley and I. Borosh, "Rivest-Shamir-Adleman public-key cryptosystems do not always conceal messages," *Computers and Mathematics with Applications* **5** pp. 169-178] (1979).
- D. Coppersmith and E. Grossman, "Generators for certain alternating groups with applications to cryptography," *SIAM journal on applied mathematics* **29** pp. 624-627 (1975).
- M. Davio and J.-J. Quisquater, "Methodology in Information Security. Mutual Authentication Procedures. Application to access control.," *Proceedings 1982 Zurich International Seminar on Digital Communications*, pp. 87-92 (1982).
- Y. Desmedt, J. Vandewalle, and R. Govaerts, "Critical analysis of the Knapsack Public Key Algorithm," *IEEE Transactions on information theory*, (1982). to appear
- W. Diffie and M. E. Hellman, "New directions in cryptography," *IEEE Transactions on information theory* **IT-22** pp. 644-654 (1976).
- W. Diffie and M. E. Hellman, "Privacy and Authentication. An Introduction to Cryptography.," *IEEE Proceedings* **67**(3) pp. 397-427 (1979).
- A. M. Duguid, "Structural properties of switching networks," *Brown University Progress report*, (1959).
- A. Evans, W Kantorowitz, and E. Weiss, "A user Authentication Scheme not Requiring Secrecy in the Computer," *Communications of the ACM* **17** pp. 437-442 (1974).
- H. Feistel, "Cryptographic coding for data bank privacy," *IBM Research Report* **RC2827** (1970).
- S. W. Golomb, *Shift register sequences*, Holden Day, San Francisco (1967).
- I. J. Good, "The relationship between two Fast Fourier Transforms," *IEEE Transactions on computers* **C-20** pp. 310-317 (1971).
- E. Grossman, "Group theoretic remarks on cryptogtaphic systems based on two types of addition," IBM TJ Wattson Res. Center RC 4742 (1974).
- T. Herlestam, "Critical remarks on some public-key cryptosystems," *BIT* **18** pp. 493-496 (1978).
- E. Horowitz and S. Sahni, "Computing partitions with applications to the knapsack," *Jl of the ACM* **21** pp. 277-292 (1974).
- D. A. Huffman, "Canonical forms for information lossless finite state logical machines," *IRE Transactions on circuit theory* **CT-6** pp. 41-59 (1959). Special supplement
- I. Ingemarson, "A user authentication scheme based on a trapdoor one-way function," IBM Res. Rpt (1980).
- I. Ingemarsson and C. K. Wong, "A conference Key Distribution System," IBM Research Report RC 8236 (#35599) (1980).

. J. B. Kam and G. I. Davida, "Structured design of substitution-permutation encryption networks," *IEEE Transactions on computers* **C-28** pp. 747-753 (1979).

. A. G. Konheim, *Cryptography: A Primer*, J Wiley, New York (1981).

. L. Lowenheim, "Gebietdeterminanten," *Math. Ann* **79** pp. 222-236 (1919).

. R. McEliece, "A public key cryptosystem based on algabraic theory," Deep space network Progr. Rpt JPL., Pasadena (1978).

. R. C. Merkle and M. E. Hellman, "Hiding information and signatures in trapdoor knapsacks," *IEEE transactions on information theory* **24** pp. 525-530 (1978).

. R. C. Merkle, "Protocols for Public-Key Cryptosystems," *Proc. 1980 Conference on Security and Privacy. IEEE. N.Y.*, pp. 122-134 (1980).

. R. Morris, N. J. A. Sloane, and A. D. Wyner, "Assessment of the NBS proposed Data Encryption Standard," *Cryptologia* **1** pp. 301-306 (1977).

. M. A. Morrison and J. Brillhart, "A method for factoring and the factorization of F7," *Math. Comp.* **29** pp. 183-205 (1975).

. V. J. Neiman, "Structure et commande optimales des reseaux de connexion sans bloquage," *Annales des telecommunications* **24** pp. 232-238 (1969).

. J. H. Pollard, "A Monte-Carlo Method for Factorization," *BIT* **15** pp. 331-334 (1975).

. R. L. Rivest, A. Shamir, and L. Adleman, "A Method for Obtaining Digital Signatures and Public-Key Cryptosystems," *Communications of the ACM* **21**(2) pp. 120-126 (1978).

. R. L. Rivest, "Remarks on a proposed cryptanalytic attack on the MIT public-key cryptosystem," *Cryptologia*, pp. 62-65 (1978).

. R. L. Rivest, "" "Critical remarks on "Critical Remarks on some public-key cryptosystems" ," *BIT* **19** pp. 274-275 (1979).

. C. Ronse, "Substitution networks," *Philips Research Laboratory. Brussels* **R-444** (1980).

. S. Rudeanu, *Boolean functions and equations*, North Holland, Amsterdam (1974).

. A. Shamir, "On the Power of Commutativity in Cryptography," pp. 582-595 in *Automata, Languages and Programming. ICALP_80 Lecture Notes*, Springer, Berlin (1980).

. A. Shamir and R. E. Zippel, "On the security of the Merkle-Hellman cryptographic scheme," *IEEE transactions on information theory* **IT-26** pp. 339-340 (1980).

. A. Shamir, *A polynomial time algorithm for breaking Merkle-Hellman cryptosystems*, The Neiman Insititute, Rehovot, Israel (1982). Research announcement; preliminary draft

. C. E. Shannon, "Communication theory of secrecy systems," *BSTJ* **28** pp. 656-715 (1949).

. G. J. Simmons and M. J. Norris, "Preliminary comments on the MIT public-key cryptosystem," *Cryptologia* **1** (4) pp. 406-414 (1977).

. G. J. Simmons, "A System for Point-of-Sale or Access User Authentication and Identification," *IEEE Workshop on Communication Security*, (1981).

. D. Slepian, "Two theorems on a particular switching network," *Unpublished manuscript*, (1952).

. R. Solovay and V. Strassen, "A fast Monte-Carlo test for primality," *SIAM Jl. of computing* **6** pp. 84-85 (1977).

. N. T. Tsao-Wu, "On Neiman's algorithm for the control of rearrangeable switching networks," *IEEE transactions on communications* **COM-22** pp. 737-742 (1974).

. A. Waksman, "A permutation network," *Jl ACM* **15** pp. 159-163 (1968).

. A. M. Whitehead, "Memoir on the algebra of symbolic logic," *Amer. Jl of Math* **23** pp. 139-165; 297-316 (1901).

. H. C. Williams and B. Schmid, "Some remarks concerning the MIT public-key cryptosystem," *BIT* **19** pp. 525-538 (1979).

SIMPLE SUBSTITUTION CIPHERS

Andrea Sgarro
Università di Trieste

1. Introduction

Substitution ciphers are a rather unsophisticated and time-honoured class of ciphers, to the extent that one may feel that they appeal to aesthetical-minded mathematicians rather than to communications engineers or to computer scientists. However, one should not overlook the fact that individually weak ciphers can be combined into a network to give a hopefully strong cryptographic system: an obvious example here is the Data Encryption Standard, or DES, developed by IBM, which is built up by concatenating substitution and transposition ciphers.

A description of substitution ciphers for single letters, or simple substitution ciphers follows. Consider a stationary memoryless source with alphabet $\mathscr{A} = \{a_1, a_2, \ldots, a_s\}$, $s \geq 2$, ruled by the probability distribution, or p.d., $P = \{p_1, p_2, \ldots, p_s\}$ with non-zero components. Only an insecure (noiseless) channel is available to transmit the confidential information output by the source. Security is furthered in the following way: out of $\mathscr{K}$, the set of all the s! permutations of the source alphabet, a permutation is randomly chosen to serve as a key; more precisely, the random choice of the key is uniform over $\mathscr{K}$ and independent of the source outputs. The key is chosen once for all and communicated to the legitimate receiver by means of a secure (but presumably too costly, or time-consuming, etc.) "special" channel. Letters output by the source are changed using the key and the resulting encrypted information is sent over the "normal" insecure channel, where it is intercepted by a wire-tapper.

As for an assessment of the security provided by the cipher, we make the pessimistic but standard assumption that the wire-tapper knows the statistics of the source and the type of cipher used; he ignores instead which message has been output by the source and which key has been selected by the legitimate users. Let us take the point of view of the wire-tapper after intercepting the cryptogram (the enciphered message string). If n, the length of the string, is sufficiently large the law of large numbers allows decryption, at least with high probability, whenever the components of P are distinct because the original message letters are identified by their frequency; instead, if in P there are ties there will be a certain amount of "uncertainty" left, however long the output string may be. It is already apparent that a simple substitution cipher is safe to use only when at least one of these rather unpracticable conditions is met: the message length is small, and/or there are enough ties in P.

A more careful analysis will be performed below for "long" strings (results will be asymptotic). The performance of the cipher will be assessed according to two criteria: either by evaluating the equivocation or average uncertainty on side of the wire-tapper, or by evaluating the probability that he makes an error when using the best statistical estimation rule to decrypt the intercepted cryptogram or to identify the key used (the latter task is more difficult when two or more source letters do not occur in the message string).

A classical reference for substitution ciphers and for cryptography in general is the famous paper by Shannon [1], where the meaning of equivocation is discussed at large. Simple substitution ciphers have been considered more or less recently in papers published on the IEEE Transactions on Information Theory ([2], [3], which tackle the equivocation approach, and [4], which tackles the error-probability approach). The technique used in [4] is that of "exact types" or "composition classes" made popular in a recent book on Information Theory [5]. Below we present all our material using the exact-type technique. Although this compact presentation is novel in some points, we can make no claim to new results.

Section 2 contains mathematical preliminaries. Section 3 deals with the equivocation approach, while section 4 deals with the error-probability approach. Results are commented upon in a final remark.

2. Mathematical preliminaries

Consider the space of all probability distributions (p.d.'s) or probability vectors over the alphabet $\mathcal{A} = \{a_1, a_2, \ldots, a_s\}$. Let $P = \{p_1, p_2, \ldots, p_s\}$ be a p.d. with strictly positive components and let $Q = \{q_1, q_2, \ldots, q_s\}$ be any other p.d.; their (informational) *divergence* $D(Q\|P)$ is defined as

$$D(Q\|P) = \sum_{i=1}^{s} q_i \log \frac{q_i}{p_i}$$

(forms like 0 log 0 are meant to be zero; logs and exps are taken to any base greater than 1, e.g. 2).

Consider the s^n sequences of $\mathscr{A}^n$, the n-th Cartesian power of $\mathscr{A}$. A *composition class* (c.c.) or an *exact type* on $\mathscr{A}^n$ is a subset $\mathscr{C}$ made up of a sequence $\underline{a}$ and of all its permutations. A c.c. $\mathscr{C}$ on $\mathscr{A}^n$ is identified by its *composition* $(n_1, n_2, \ldots, n_s)$, where n_i is the number of occurrences of letter a_i in any of its sequences. If the components of the composition of $\mathscr{C}$ are divided by n one obtains a p.d. F on $\mathscr{A}$ which is called the *empirical distribution* (e.d.) associated to $\mathscr{C}$. The length n is called the *order* of the c.c. $\mathscr{C}$ and of the corresponding e.d. F (of course any multiple of n is still an order of F).

Details on c.c.'s can be found, e.g., in [5]. We will list some basic facts; below $\mathscr{C}$ is a c.c. of order n, F is its associated e.d. and $\underline{a}$ belongs to $\mathscr{C}$; H(·) denotes entropy.

$$D(Q\|P) \geq 0, \quad D(Q\|P) = 0 \quad \text{iff} \quad P = Q \tag{1}$$

the number of c.c.'s of order n is not greater than $(n+1)^s$ (2)

$$P^n(\underline{a}) = \exp\{-n[H(F) + D(F\|P)]\} \tag{3}$$

$$(n+1)^{-s} \exp\{-n\, D(F\|P)\} \leq P^n(\mathscr{C}) \leq \exp\{-n\, D(F\|P)\} \tag{4}$$

We recall also that entropy is strictly concave and that divergence is strictly convex in both arguments. The following obvious lemma is stated explicitly for ease of reference.

Lemma 1. Let $\mathscr{D}$ be a domain of p.d.'s over $\mathscr{A}$ and let $\mathscr{D}_n$ be the subset of $\mathscr{D}$ made up by all its e.d.'s of order n. Then

$$\lim_n \min_{Q\in\mathscr{D}_n} D(Q\|P) = \min_{Q\in\mathscr{D}} D(Q\|P) .$$

By a domain we mean the closure of a non-empty open set of p.d.'s (closure is taken in the usual topology of real vectors).

The lemma below is a useful inequality due to Blom [2]; X and Y are finite-range random variables; summations are extended to all values taken by the variables; $H(Y \mid X)$ is the conditional entropy of Y given X.

Lemma 2.

$$H(Y|X) \leqslant \log \sum_i \sum_j \sum_k \sqrt{\Pr\{X=x_i,\ Y=y_j\}\ \Pr\{X=x_i,\ Y=y_k\}}$$

Proof.

Set $p_i = \Pr\{X=x_i\}$, $p_{j|i} = \Pr\{Y=y_j|X=x_i\}$, $p_{ij} = p_i\ p_{j|i}$. Then

$$H(Y|X) = \sum_i p_i\ H(Y|X=x_i) = \sum_i p_i [2 \sum_j p_{j|i} \log (1/\sqrt{p_{j|i}})]$$

$$= \sum_i p_i [2 \log \prod_j (1/\sqrt{p_{j|i}})^{p_{j|i}}] \leqslant \sum_i p_i \log (\sum_j p_{j|i}/\sqrt{p_{j|i}})^2$$

$$= \sum_i p_i \log \sum_j \sum_k \sqrt{p_{j|i}\ p_{k|i}} \leqslant \log \sum_i p_i \sum_j \sum_k \sqrt{p_{j|i}\ p_{k|i}}$$

$$= \log \sum_i \sum_j \sum_k \sqrt{p_{ij}\ p_{ik}}$$

The two inequalities hold because the geometric mean is not larger than the arithmetic mean and because the logarithm is a concave function, respectively. QED

Three parameters depending on P are needed. Assume that in P the distinct components are d $(1 \leqslant d \leqslant s)$ and that they occur $s_1, s_2, \ldots, s_d$ times each, respectively; $s_1 + s_2 + \ldots + s_d = s$. Set

(5) $$A = A(P) = s_1!\ s_2!\ \ldots\ s_d!$$

Of course $1 \leqslant A \leqslant s!$; $A = 1$ iff in P there are no equal components; $A = s!$ iff P is uniform. Then set

(6) $$B = B(P) = \min_{i,j:\ p_i \neq p_j} (\sqrt{p_i} - \sqrt{p_j})^2$$

for P non-uniform. Finally set

(7) $$D = D(P) = \min_{i \neq j} (p_i + p_j).$$

Heuristically, it is often useful to see the divergence as a sort of "non-metric distance" between p.d.'s. We now solve a few problems relative to the resulting "geometry".

We tackle the first "geometric" problem.

In the space of all p.d.'s over $\mathscr{A}$ consider those p.d.'s which are constrained to be zero in correspondence to letters outside a (proper) subalphabet $\mathscr{B} \subset \mathscr{A}$. Fix P as above. We wish to find the constrained p.d. U which is nearest to P in the sense of divergence; set $\alpha = \sum_{\mathscr{B}} p_i$.

$$D(U\|P) = \sum_{\mathscr{B}} u_i \log \frac{u_i}{p_i} = -\log \alpha + \sum_{\mathscr{B}} u_i \log \frac{\alpha\, u_i}{p_i}$$

$$\geqslant -\log \alpha = -\log \sum_{\mathscr{B}} p_i \,.$$

The inequality follows from (1) because the summation is a divergence over the subalphabet $\mathscr{B}$; the condition for equality is also given in (1). The constrained p.d. nearest to P is the p.d. U whose unconstrained components are proportional to the corresponding components of P. We are now in a position to obtain the following lemma.

Lemma 3 (for $s \geqslant 3$).

Let $\mathscr{A}_n^*$ be the set of n-tuples in $\mathscr{A}^n$ such that two or more letters do not occur in them. Then

$$\lim n^{-1} \log P^n(\mathscr{A}_n^*) = \log (1 - D(P))$$

Proof.

$\mathscr{A}_n^*$ is a union of composition classes or c.c.'s. If $\mathscr{C}_n$ is a c.c. of maximal probability in the union, (2) implies

$$P^n(\mathscr{C}_n) \leqslant P^n(\mathscr{A}_n^*) \leqslant (n+1)^s\, P^n(\mathscr{C}_n)$$

and therefore

$$\lim n^{-1} \log P^n(\mathscr{A}_n^*) = \lim n^{-1} \log P^n(\mathscr{C}_n)$$

provided that the limits exist. On the other hand, cf. (4):

$$(n+1)^{-s} \exp\{-n\, D(F_n\|P)\} \leqslant P^n(\mathscr{C}_n) \leqslant \exp\{-n\, D(F_n\|P)\}$$

where F_n is the e.d. corresponding to $\mathscr{C}_n$. The bounds are weakened by taking the e.d. which minimizes $D(U\|P)$ over the set of all e.d.'s of order n which have at least two null components. Asymptotically the constraint that U be empirical can be dropped (lemma 1) and lemma 3 is obtained using the "geometric" result found above. QED

We tackle two more "geometric" problems (P is fixed as above and for the moment Q is any other p.d.):

a) minimize $D(U\|P) + D(U\|Q)$, U unconstrained,

b) minimize $D(U\|P)$ with the constraint $D(U\|Q) \leqslant D(U\|P)$.

Problem a) is easy to solve using aaain (1):

$$D(U\|P) + D(U\|Q) = \sum_i u_i \log \frac{u_i^2}{p_i q_i} = 2 \sum_i u_i \log \frac{u_i}{\sqrt{p_i q_i}}$$

$$\geqslant - 2 \log \sum_i \sqrt{p_i q_i}$$

with equality if and only if $u_i \propto \sqrt{p_i q_i}$ ($\propto$ means "proportional to").

Problem b) is rather more complex and we shall solve it only in a particular case. First we show that the constraint can be changed to $D(U\|P) = D(U\|Q)$. Fix a number α, $0 < \alpha < D(P\|Q)$. Assume that U minimizes $D(U\|P)$ with the constraint $D(U\|Q) \leqslant \alpha$; we claim that U is a border p.d., that is that $D(U\|Q) = \alpha$. An inverse proof of this claim is obtained by observing that any point V on the open Euclidean segment (U, P) satisfies $D(V\|P) < D(U\|P)$ because of strict convexity; for V sufficiently near to U $D(V\|Q) \leqslant \alpha$ still holds because of continuity.

Let us go back to problem a). Whenever the p.d. U which solves problem a) is such that $D(U\|P) = D(U\|Q)$, problems a) and b) have clearly the same solution. An instance of this is the case when Q is obtained from P by a single exchange of distinct components (we assume also that P is not uniform): then $D(U\|P) = D(U\|Q)$ holds because the two components of U corresponding to the exchanged components of P are equal to each other.

In lemma 4 below we need p.d.'s Q which are permutations of P. It seems that we have proved more than needed for problem a) but less than needed for problem b). However, we shall find a way out. Note that, unlike in [4], to prove lemma 4 we can dispense with using Lagrange multipliers.

Lemma 4 (for P non-uniform)

Let $\mathcal{E}$ be the set of p.d.'s, P excluded, which are obtained by permuting the components of P. Assume that p_{i_1} and p_{i_2} yield B(P); cf. (6). Then $D(U\|P) + D(U\|Q)$ is minimized over $\mathcal{E}$ and over all p.d.'s U by taking $u_{i_1} = u_{i_2} \propto \sqrt{p_{i_1} p_{i_2}}$, otherwise $u_i \propto p_i$. The same U solves the problem of minimizing $D(U\|P)$ with the constraint $D(U\|Q) \leqslant D(U\|P)$ for at least a Q in $\mathcal{E}$. With this minimizing U

$$D(U\|P) = -\log[1 - B(P)].$$

Proof.

Fix Q in $\mathcal{E}$; the corresponding $U = U(Q)$ which minimizes $D(U\|P) + D(U\|Q)$ gives $D(U\|P) + D(U\|Q) = -2\log\Sigma\sqrt{p_i\, q_i}$. One has now to minimize $-\log\Sigma\sqrt{p_i\, q_i}$ over $\mathcal{E}$. One has, as already noted in [2]:

$$\max_{\mathcal{E}} \sum_i \sqrt{p_i\, q_i} = \max_{\mathcal{E}} [1 - \frac{1}{2} \sum_i (\sqrt{p_i} - \sqrt{q_i})^2] = 1 - B(P).$$

We now come to the problem of minimizing $D(U\|P)$ with the constraint $D(U\|Q) \leqslant D(U\|P)$ for at least a Q in $\mathcal{E}$. To complete the proof of the lemma the only difficulty is that $\mathcal{E}$ contains all the permutations of P distinct from P, and not only the permutations which are obtained by single exchanges. To overcome this difficulty we show that for any permutation Q_1 of P, $Q_1 \neq P$, there is a permutation Q_2 obtained from P by a single exchange of distinct components such that $D(U_2\| P) \leqslant D(U_1\| P)$; here U_1 and U_2 solve the problem of minimizing $D(U\|P)$ with the constraint $D(U\|Q_1) \leqslant D(U\|P)$ and $D(U\|Q_2) \leqslant D(U\|P)$, respectively. We proceed to a proof of this claim in three steps; to simplify notation we assume $p_1 \geqslant p_2 \geqslant \ldots \geqslant p_s$, say. Set $E_Q(P) = \min_{D(U\|Q) \leqslant D(U\|P)} D(U\|P) =$
$= \min_{D(U\|Q) = D(U\|P)} D(U\|P)$.

1st step. Let $Q = \{q_1, q_2, \ldots, q_s\}$, $R = \{q_2, q_1, q_3, \ldots, q_s\}$ and $U =$
$= \{u_1, u_2, \ldots, u_s\}$ be three p.d.'s, with $q_1 \geqslant q_2$; the components of Q and R from the third to the last coincide. We show that there is a p.d. W which is "nearer" to P than U and "nearer" to Q than U is to R ("near" is meant in the sense of divergence). Namely, if $u_1 \geqslant u_2$, take $W = U$, while if $u_1 < u_2$ take $W = \{u_2, u_1, u_3, \ldots, u_s\}$. In the first case $D(W\|P) = D(U\|P)$ and $D(W\|Q) \leqslant D(U\|R)$. In the second case $D(W\|Q) = D(U\|R)$ and $D(W\|P) \leqslant D(U\|P)$. The two inequalities can be proved by simple calculations.

2nd step. Assume now that U yields $E_R(P)$. Then $E_R(P) = D(U\|P) = D(U\|R)$. If $u_1 \geqslant u_2$, since $D(U\|Q) \leqslant D(U\|R) = D(U\|P)$, U appears in the minimization set for $E_Q(P)$; then $E_Q(P) \leqslant D(U\|P) = E_R(P)$. If $u_1 \leqslant u_2$, $D(W\|P) \leqslant D(U\|P) = D(U\|R) = D(W\|Q)$, so W appears in the minimization set for $E_P(Q)$; however $E_Q(P) = E_P(Q)$, because $E_Q(P) = \min_{D(U\|P) = D(U\|Q)} D(U\|Q) = \min_{D(U\|P) = D(U\|Q)} D(U\|P)$. Then $E_Q(P) = E_P(Q) \leqslant D(W\|Q) = D(U\|R) =$
$= E_R(P)$.

3rd step. In the first two steps we have concentrated about p_1 and p_2 ; however any couple $p_i \leqslant p_j$ might have been considered. Once a permutation R of P is given, with successive exchanges of probabilities $r_i < r_j$, $i < j$, we can bring R back to P; or rather stop at the step before the last: at every exchange E_R (P) diminishes, or at least does not increase. QED

3. Key equivocation and message equivocation

Below M_n and K denote the random message of length n and the random key; $C_n = T_K(M_n)$ is the random cryptogram, T_K denoting the random transformation corresponding to key K; $\underline{m}$, $\underline{c}$ and k, etc. are specific values of M_n, C_n, K.

We compute the key equivocation, that is the conditional entropy $H(K \mid C_n)$. We assume P non-uniform, else trivially $H(K \mid C_n) = \log s!$. One has (below $\mathcal{E}^*$ stands for $\mathcal{E} \cup \{P\}$, cf. lemma 4):

$$H(K|C_n) \overset{(i)}{=} \Sigma_k \Sigma_{\underline{c}} \Pr\{K=k,\ C_n=\underline{c}\} \log \frac{\Sigma_h \Pr\{K=h,\ C_n=\underline{c}\}}{\Pr\{K=k,\ C_n=\underline{c}\}}$$

$$\overset{(ii)}{=} \sum_k \sum_{\underline{c}} \frac{1}{s!} \Pr\{M_n = T_k^{-1}(\underline{c})\} \log \frac{\sum_h \Pr\{M_n = T_h^{-1}(\underline{c})\}}{\Pr\{M_n = T_k^{-1}(\underline{c})\}}$$

$$= \frac{1}{s!} \sum_k \sum_{\underline{c}} P^n(T_k^{-1}(\underline{c})) \log \frac{\sum_h P^n(T_h^{-1}(c))}{P^n(T_k^{-1}(\underline{c}))}$$

$$\overset{(iii)}{=} \frac{A}{s!} \sum_{\mathcal{E}^*} \sum_{\underline{m}} Q^n(\underline{m}) \log \frac{A \sum_{\mathcal{E}^*} \tilde{Q}^n(\underline{m})}{Q^n(\underline{m})} = \log A + \frac{A}{s!} \sum_{\mathcal{E}^*} \sum_{\underline{m}} Q^n(\underline{m}) \log \frac{\sum_{\mathcal{E}^*} \tilde{Q}^n(\underline{m})}{Q^n(\underline{m})}$$

$$\overset{(iv)}{=} \log A + \sum_{\underline{m}} P^n(\underline{m}) \log\left(1 + \frac{\sum_{\mathcal{E}} Q^n(\underline{m})}{P^n(\underline{m})}\right).$$

Above (i) is the definition of conditional entropy; (ii) uses the independence of message and key; in (iii) instead of transforming $\underline{m}$ we permute the components of P and recall that each permutation is obtained A times (cf. the definition (5) of A); in (iv) we use the fact that the s!/A terms in the $\mathcal{E}^*$ summation are equal. The exact expression of $H(K|C_n)$ given above grows rapidly too difficult to compute.

Now we upper- and lower-bound $H(K \mid C_n) - \log A$. To obtain the upper bound we use Blom's inequality (lemma 2).

We write back $H(K \mid C_n) - \log A$ in the form of a conditional entropy:

$$H(K|C_n) - \log A = \sum_{\underline{m}} \sum_{\mathcal{E}^*} \frac{1}{|\mathcal{E}^*|} Q^n(\underline{m}) \log \frac{\sum_{\mathcal{E}^*} \frac{1}{|\mathcal{E}^*|} Q^n(\underline{m})}{\frac{1}{|\mathcal{E}^*|} Q^n(\underline{m})}$$

$$\overset{(i)}{\leqslant} \log \sum_{\underline{m}} \sum_{Q \in \mathcal{E}^*} \sum_{\tilde{Q} \in \mathcal{E}^*} \sqrt{\frac{1}{|\mathcal{E}^*|} Q^n(\underline{m}) \frac{1}{|\mathcal{E}^*|} \tilde{Q}^n(\underline{m})}$$

$$\overset{(ii)}{=} \log \sum_{\underline{m}} \sum_{\mathcal{E}^*} \sqrt{P^n(\underline{m}) Q^n(\underline{m})} = \log (1 + \sum_{\underline{m}} \sum_{\mathcal{E}} \sqrt{P^n(\underline{m}) Q^n(\underline{m})}$$

$$\overset{(iii)}{\leqslant} \log e \sum_{\underline{m}} \sum_{\mathcal{E}} \sqrt{P^n(\underline{m}) Q^n(\underline{m})}$$

$$\overset{(iv)}{=} \log e \sum_{F_n} |\mathcal{C}_{F_n}| \sum_{\mathcal{E}} \sqrt{\exp\{-n[2H(F_n) + D(F_n \| P) + D(F_n \| Q)]\}}$$

$$\overset{(v)}{\leqslant} \log e \sum_{F_n} \sum_{\mathcal{E}} \sqrt{\exp\{-n[D(F_n \| P) + D(F_n \| Q)]}$$

$$\overset{(vi)}{\leqslant} \log e\ (n+1)^s\ |\mathcal{E}| \sqrt{\exp\{-n \min_{F_n} \min_{\mathcal{E}} [D(U \| P) + D(U \| Q)]\}}$$

Above (i) is Blom's inequality; in (ii) we have used the fact that the same term is summed $|\mathcal{E}^*|$ times (the vertical bars denote cardinality); (iii) is the obvious inequality $\log x \leqslant (x-1) \log e$; in (iv) we have pooled together all messages which are in the same c.c. $\mathcal{C}_{F_n}$ of e.d. F_n and have used (3); (v) follows from the inequality $|\mathcal{C}_{F_n}| \leqslant \exp\{n H(F_n)\}$ (use (3) and (4)); (vi) follows from (2).

Lemma 1 allows us to make use of lemma 4, so that

$$H(K|C_n) \leqslant \log A + \exp\{n[\log(1 - B(P)) + \delta_n^{(1)}]\}, \quad \lim_n \delta_n^{(1)} = 0.$$

Now we go to the upper bound. In the passages below we assume that $\tilde{Q}$ is a p.d. in $\mathcal{E}$ obtained by exchanging the probabilities which yield B(P) (any two in case of ties):

$$H(K|C_n) - \log A \geqslant \sum_{\underline{m}} P^n(\underline{m}) \log [1 + \frac{\tilde{Q}^n(\underline{m})}{P^n(\underline{m})}$$

$$\geqslant P^n(\mathscr{C}_{U_n}) \log [1 + \frac{\tilde{Q}^n(\mathscr{C}_{U_n})}{P^n(\mathscr{C}_{U_n})}$$

In the second inequality we have retained only messages whose associated e.d. U_n has been chosen as to converge to the p.d. U which solves the minimization problem of lemma 4 (any such U in case of ties); we also assume that the components of U_n corresponding to the exchanged components of P are equal, the same being true for their limits. Then $D(U_n \| P) = D(U_n \| Q)$ and

$$H(K|C_n) \geqslant P^n(\mathscr{C}_{U_n}) \log 2 = P^n(\mathscr{C}_{U_n})$$

$\mathscr{C}_{U_n}$ being the c.c. corresponding to U_n. The probability of this c.c. is lower-bounded using (4); therefore, by lemma 1:

$$H(K|C_n) \geqslant \log A + \exp\{n[\log(1 - B(P)) + \delta_n^{(2)}]\}, \ \lim_n \delta_n^{(2)} = 0$$

Recalling the upper bound we have the expression

(8) $$H(K|C_n) = \log A + \exp\{n[\log(1 - B(P)) + \varepsilon_n^{(1)}]\}, \ \lim_n \varepsilon_n^{(1)} = 0.$$

We now turn to message equivocation, that is to the conditional entropy $H(M_n \mid C_n)$; we assume $s \geqslant 3$, else $H(M_n|C_n) = H(K|C_n)$.

One has $H(K, M_n \mid C_n) = H(K \mid C_n) + H(M_n \mid K, C_n) = H(M_n \mid C_n) + H(K \mid M_n, C_n)$. Since $H(M_n \mid K, C_n) = 0$, one has $H(M_n \mid C_n) = H(K \mid C_n) - H(K \mid M_n, C_n)$; therefore it will be enough to compute $H(K \mid M_n, C_n)$. This conditional entropy is itself an equivocation which has its own interest in case of a so-called "plain-text attack", when the wire-tapper knows both the message and the cryptogram and wants to identify the key for future use. Below $v(\underline{m})$ denotes the number of letters which do not occur in $\underline{m}$; $0 \leqslant v(\underline{m}) \leqslant s - 1$.

$$H(K|M_n, C_n) = \sum_{\underline{m}} \sum_{\underline{c}} \Pr\{M_n = \underline{m}, C_n = \underline{c}\} \, H(K|M_n = \underline{m}, C_n = \underline{c})$$

$$= \sum_{\underline{m}} \log v(\underline{m})! \sum_{\underline{c}} \Pr\{M_n = \underline{m}, C_n = \underline{c}\} = \sum_{\underline{m}} \log v(\underline{m})! \ \Pr\{M_n = \underline{m}\}$$

Above terms like $\Pr\{M_n = \underline{m}, C_n = \underline{c}\}\ H(K \mid M_n = \underline{m}, C_n = \underline{c})$ are meant to be zero whenever $\Pr\{M_n = \underline{m}, C_n = \underline{c}\}$ is zero. On the other hand, since $\log v(\underline{m})!$ is zero for $v(\underline{m}) = 0$ and $v(\underline{m}) = 1$, we may keep in the last summation only messages which belong to $\mathscr{A}_n^*$ say, where $\mathscr{A}_n^*$ contains all messages where at least two letters are missing ($v(\underline{m}) \geq 2$). Then

$$\log 2!\ \ P^n(\mathscr{A}_n^*) \leq H(K|M_n, C_n) \leq \log(s-1)!\ \ P^n(\mathscr{A}_n^*)$$

The behaviour of $P^n(\mathscr{A}_n^*)$ is already known (cf. lemma 3). Therefore

$$H(K|M_n, C_n) = \exp\{n[\log(1 - D(P)) + \varepsilon_n^{(2)}]\},\ \lim_n \varepsilon_n^{(2)} = 0\,.$$

This, together with (8), gives the following theorem.

Theorem 1 (on equivocations).

For the stationary memoryless source and the simple substitution cipher described above one has

$$H(K|C_n) = \log A + \exp\{n[\log(1 - B(P)) + \varepsilon_n^{(1)}]\},\ \lim_n \varepsilon_n^{(1)} = 0$$

(the exponential term is dropped for P uniform), and

$$H(M_n|C_n) = H(K|C_n) - \exp\{n[\log(1 - D(P)) + \varepsilon_n^{(2)}]\},\ \lim_n \varepsilon_n^{(2)} = 0$$

(the exponential term is dropped for $s = 2$).

Corollary (P non-uniform and $s \geq 3$).

If $B(P) < D(P)$

$$H(M_n|C_n) = \log A + \exp\{n[\log(1 - B(P)) + \varepsilon_n^{(3)}]\},\ \lim_n \varepsilon_n^{(3)} = 0.$$

if $B(P) > D(P)$

$$H(M_n|C_n) = \log A - \exp\{n[\log(1 - D(P)) + \varepsilon_n^{(4)}]\};\ \lim_n \varepsilon_n^{(4)} = 0\,.$$

The restrictions in the corollary have been assumed only to avoid trivial specifications. The corollary is easily proved; e.g., for $B < D$, it is enough to use the bounds ·

$$\frac{1}{2}\exp\{n[\log(1 - B) + \varepsilon_n^{(1)}]\} \leq \exp\{n[\log(1 - B) + \varepsilon_n^{(1)}]\} - \exp\{n[\log(1 - D) + \varepsilon_n^{(2)}]$$

$\leq \exp\{n[\log(1-B) + \varepsilon_n^{(1)}]\}$, which hold for n large enough.

The case $B > D$ may hold true only when the components of P yielding D are equal (this in turn implies that the constant term log A is positive); in fact, as observed in [3], if p_1 and p_2 , say, yield D and if $p_1 \neq p_2$, then $B \leq (\sqrt{p_1} - \sqrt{p_2})^2 < p_1 + p_2$.

4. The probability-of-error approach

The parameters A(P), B(P) and D(P) turn up also in the probability-of-error approach. Below we consider only key identification in detail. Roughly speaking, message identification bears the same relation to key identification as message equivocation does to key equivocation; we shall quote the main results for message identification without proofs; for these we refer to [4].

Since in our model prior probabilities are given, it will be generally agreed that the best statistical procedure on side of the wire-tapper for identifying the key is the Bayesian procedure of maximizing posterior probabilities: the wire-tapper decides for a key whose conditional probability given the intercepted cryptogram is maximal. Other procedures are however equivalent to the Bayesian one, as proved in the following lemma:

Lemma 5.

The following procedures are equivalent:

a) maximize the conditional key probability given the cryptogram (Bayesian approach);
b) maximize the conditional cryptogram probability given the key (maximum likelihood approach);
c) maximize the unconditional message probability over the set of messages which have positive conditional probability given the cryptogram; then pick any key which applied to the selected message gives the cryptogram.

(In the maximizations of a), b) and c) ties are resolved arbitrarily; note also that above the cryptogram is a constant).

Proof.

The output message has the same "form" as the cryptogram, since there is a key which transforms the message into the cryptogram. One has:

$$\Pr\{K = k | C_n = \underline{c}\} \propto \Pr\{K = k\} \Pr\{C_n = \underline{c} | K = k\} \propto \Pr\{C_n = \underline{c} | K = k\}$$

$$= \Pr\{M_n = T_k^{-1}(\underline{c}) | K = k\} = \Pr\{M_n = T_k^{-1}(\underline{c})\}$$

The first passage is Bayes' rule; in the second we use the fact that the key probability is a constant and so prove the equivalence of a) and b); in the last we use the independence of key and message. The conditional probability of the message given the cryptogram is positive iff there is a key which transforms the message into the cryptogram. QED

We now describe the key identification procedure in less formal terms.

From the enemy's standpoint the cryptogram is ouput by a source whose statistics he does not know, since it is determined not only by the well-known message statistics, but also by the unknown key. Each key corresponds to a possible "explanation" of the observed cryptogram; if there are ties in the components of P, however, some of these explanations are statistically undistinguishable or statistically equivalent. If we lump together equivalent keys into a single statistical hypothesis, what the enemy is left with is a normal problem of hypothesis discrimination; each hypothesis corresponds to a probability distribution obtained by permuting the components of P. The enemy adopts a maximum-likelihood discrimination criterion (cf. lemma 5): he chooses a permutation of P such as to maximize the probability of his observation, that is the probability of the intercepted cryptogram.

Choosing a permutation of P is the same as choosing a key only if the components of P are all distinct. Otherwise the key identification procedure is not yet terminated and a whole class of equivalent keys is still there; a single key is then chosen according to an arbitrary rule (obviously any rule is equally good, or rather equally bad, at this stage, because from the wire-tapper's standpoint the equivalent keys obtained through maximum-likelihood discrimination are all equally likely; as a matter of fact even at the first stage there might have been classes of keys giving maximal probabilities and also in this case the choice of a single class is made according to an arbitrary rule; however it will be shown that the latter ambiguity is asymptotically irrelevant).

We now turn to computations: $P_{e,k} = P_{e,k}(n)$ denotes the probability of incorrect key identification. We dispense with the trivial case when P is uniform: then $P_{e,k} = (s! - 1)/s! = (A-1)/A$ (A has been defined in (5)).

Below we prove that for P non-uniform

$$P_{e,k} = \frac{A-1}{A} + \exp\{n[\log(1-B) + \varepsilon_n^{(5)}]\}, \quad \lim_n \varepsilon_n^{(5)} = 0. \tag{9}$$

Denote by $P_{e,k}^{(1)} = P_{e,k}^{(1)}(n)$ the probability of error in the first part of the key identification procedure (key-class identification; see above).

Write $P_{e,k}$ as

$$P_{e,k} = P_{e,k}^{(1)} + (1 - P_{e,k}^{(1)})\,\frac{A-1}{A} = \frac{A-1}{A} + \frac{1}{A}\,P_{e,k}^{(1)}\ ; \tag{10}$$

it is obvious that $(A-1)/A$ is the probability of error in the whole key-identification procedure subject to the condition that the first part has been successful; cf. definition (5).

Let us turn to $P^{(1)}_{e,k}$. Since P is not uniform the set $\mathcal{E}$ which contains the permutations of P distinct from P is not void. The enemy will be wrong whenever there is a permutation Q in $\mathcal{E}$ such that the Q-probability of the message is greater than its P-probability; while he will be right if the P-probability of the message is greater than any of its Q-probabilities, $Q \in \mathcal{E}$; cf. point c) in lemma 5. Because of the possibility of ties we can only give upper and lower bounds for $P^{(1)}_{e,k}$

$$\sum_{\underline{m}\in\mathcal{H}} P^n(\underline{m}) \leqslant P^{(1)}_{e,k} \leqslant \sum_{\underline{m}\in\mathcal{H}'} P^n(\underline{m})$$

where $\mathcal{H}$ contains all messages $\underline{m}$ such that $P^n(\underline{m}) < Q^n(\underline{m})$ for some Q in $\mathcal{E}$ and $\mathcal{H}'$, contains all messages $\underline{m}$ such that $P^n(\underline{m}) \leqslant Q^n(\underline{m})$ for some Q in $\mathcal{E}$; $\mathcal{H} \subseteq \mathcal{H}$!

Since $\mathcal{H}$ and $\mathcal{H}'$ are obviously unions of c.c.'s $\mathscr{C}_j$, the bounds can be rewritten as:

$$(11) \qquad \sum_{j\in\mathcal{J}} P^n(\mathscr{C}_j) \leqslant P^{(1)}_{e,k} \leqslant \sum_{j\in\mathcal{J}'} P^n(\mathscr{C}_j)$$

where $\mathcal{J}$ and $\mathcal{J}'(\mathcal{J}\subseteq\mathcal{J}')$ are suitable sets of indices. More precisely, using (c), $j \in \mathcal{J}$ iff $D(F_j \| Q) < D(F_j \| P)$ for some Q in $\mathcal{E}$, while $j \in \mathcal{J}'$ iff $D(F_j \| Q) \leqslant D(F_j \| P)$ for some Q in $\mathcal{E}$; F_j is the e.d. corresponding to the c.c. $\mathscr{C}_j$.

By (2) and (4), (11) can be weakened to:

$$(12) \qquad (n+1)^{-s} \exp\{-n \min_{j\in\mathcal{J}} D(F_j \| P)\} \leqslant P^{(1)}_{e,k} \leqslant (n+1)^{s} \exp\{-n \min_{j\in\mathcal{J}'} D(F_j \| P)\}$$

Take n^{-1} log of the three sides; lemma 1 and lemma 4 give

$$\lim_n [-n^{-1} \log P^{(1)}_{e,k}(n)] = \min_{U\in\mathcal{J}_2} D(U\|P) = -\log(1-B) ;$$

above $\mathcal{J}_2$ is the set of all p.d.'s U over $\mathscr{A}$ such that $D(U \| Q) \leqslant D(U\|P)$ for at least one Q in $\mathcal{E}$; we have also used the fact that $\inf_{U\in\mathcal{J}_3} D(U\|) = \min_{U\in\mathcal{J}_2} D(U\|P)$, $\mathcal{J}_3$ being the set of all p.d.'s U such that $D(U\|Q) < D(U\|P)$ for at least a Q in $\mathcal{E}$.

In (10) one has also $1/A \; P^{(1)}_{e,k}$; obviously

$$\lim_n n^{-1} \log \frac{1}{A} P^{(1)}_{e,k}(n) = \lim_n n^{-1} \log P^{(1)}_{e,k}(n) = \log(1-B) .$$

This terminates the proof of (9).

In theorem 2 below we state also the results relative to message identification;

$P_{e,m} = P_{e,m}(n)$ denotes the probability that the message-identification procedure is incorrect; for $s \geq 3$ $P_{e,m} < P_{e,k}$ because it is possible that two or more letters do not occur in the message. As for the optimal message-identification procedure, in [4] it has been proved that the best procedure in the Bayesian sense is obtained by simply applying the inverse key obtained as explained above to the intercepted cryptogram.

Theorem 2 (on probabilities of error).

For the stationary memoryless source and the simple substitution cipher described above one has

$$P_{e,k}(n) = \frac{A-1}{A} + \exp\{n[\log(1-B(P)) + \varepsilon_n^{(5)}]\}, \quad \lim_n \varepsilon_n^{(5)} = 0,$$

(the exponential term is dropped for P uniform), and

$$P_{e,m}(n) = P_{e,k}(n) - \exp\{n[\log(1-D(P)) + \varepsilon_n^{(6)}]\}, \quad \lim_n \varepsilon_n^{(6)} = 0$$

(the exponential term is dropped for $s = 2$).

Corollary (P non-uniform and $s \geq 3$).

If $B(P) < D(P)$

$$P_{e,m}(n) = \frac{A-1}{A} + \exp\{n[\log(1-B(P)) + \varepsilon_n^{(7)}]\}, \quad \lim_n \varepsilon_n^{(7)} = 0;$$

if $B(P) > D(P)$

$$P_{e,m}(n) = \frac{A-1}{A} - \exp\{n[\log(1-D(P)) + \varepsilon_n^{(8)}]\}, \quad \lim_n \varepsilon_n^{(8)} = 0.$$

Again the restrictions in the corollary serve only to avoid trivial specifications. The kinship of these results to those obtained for the equivocation approach is striking. We mention the fact that finite-length bounds for equivocations and error probabilities are easily obtained by slightly deepening the technique based on composition classes or exact types that we have used (see [4], where this has been done for the error probabilities).

5. Final remarks

The safety of a simple substitution cipher has been assessed using two criteria, equivocation and error probability on side of the wire-tapper. The results obtained, however, are strictly related. Let us first consider the case when the wire-tapper is interested in the key used rather than in the message sent. Then both the equivocation and the error probability are the sum of two terms. The first term is constant (does not decrease with message length) and depends on A, that is on the number of ties in the components of the source statistics; it is zero only when there are no ties and has its maximum value (log s! and (s! − 1)/s! respectively) for P uniform, when the wire-tapper has actually to "guess" the information he needs, the intercepted cryptogram being of no help to him. Letters of the same probability cannot be discriminated, so that the constant term corresponds to "unremovable uncertainty". When P is not uniform an exponentially decreasing term has to be summed to the first term; asymptotically this exponential term depends only on B, that is on the two distinct probabilities in the source statistics whose square roots have the smallest distance (two letters with these probabilities are the most difficult to discriminate). Let us now assume that the wire-tapper is interested in the message sent and that the source alphabet has at least 3 letters; then the key cannot be computed from the couple message-cryptogram whenever two or more letters do not occur in the message, so that both equivocation and error probability decrease with respect to the case when the wire-tapper is interested in the key. Computations show that an exponentially decreasing term has to be subtracted from the values found for the key; asymptotically this term depends only on D, that is on the two smallest (not necessarily distinct) probabilities in the source statistics (letters with these probabilities are "often" missing in the message output by the source). Usually the positive exponential term dominates the negative one, at least asymptotically, so that identifying the key or message has "approximately" the same difficulty. However, when the sum of the two smallest probabilities is less than the difference of squares mentioned above, it is the negative term which dominates the positive one. In this exceptional case, whose occurrence implies that the two smallest probabilities coincide, one obtains for equivocation or for error-probability a value which is below that corresponding to "unremovable uncertainty" for the key.

REFERENCES

[1] C. Shannon, "Communication theory of secrecy systems", Bell Syst. Tech. J., vol. 28, pp. 656-715, Oct. 1949.

[2] R.J. Blom, "Bounds on key equivocation for simple substitution ciphers", IEEE Trans. Inform. Theory, vol. IT-25, pp. 8-18, Jan. 1979.

[3] J.G. Dunham, "Bounds on message equivocation for simple substitution ciphers", IEEE Trans. Inform. Theory, vol. IT-26, pp. 522-527, Sept. 1980.

[4] A. Sgarro, "Error probabilities for simple substitution ciphers" IEEE Trans. Inform. Theory, vol. IT-29, March 1983.

[5] I. Csiszár and J. Körner "Information theory: coding theorems for discrete memoryless systems", New York, Academic, 1981.

STREAM CIPHERS

Thomas Beth
Universität Erlangen-Nürnberg

1. Introduction:

In any communication system the users face the problem of protecting the transmitted data against a loss of their - objective or subjective - quality.

Objective damages are those which destroy or disturb the data, e.g. due to fading in radio transmission, lightening burst on telephone lines, magnetization faults on tape, systematic jamming of frequency bands etc.

While the damages generally can be recognized by the users of the system, the second kind - the loss of subjective quality - is mainly described by the phenomenon that it normally occurs without the user's detecting it.

Examples for this are interception (passive violation)

Picture 1: passive violation

or <u>forgery</u> <u>(active violation)</u>

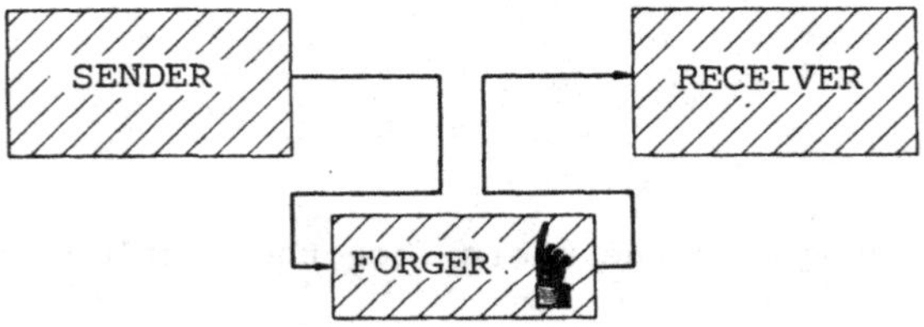

Picture 2: active violation

Protection against the first kind - the objective loss - is provided by the methods of filtering, error-correcting codes (cf. Blahut's contribution in this volume), the recent techniques of frequency hopping and spread spectra (cf. Pursley's and Ephremides' papers in this volume). The two main types of subjective quality loss, interception and forgery, are usually prohibited by means of cryptographic ciphers.

The papers by Davio/Goethals and Nemetz (also in this volume) give another introduction of how to scramble data in such a way that any violation can be excluded since an enemy (interceptor, forger) would have to understand the plaintext version first. In the sequel we shall present the corresponding countermeasures.

2. Definitions and Notations:

The following elementary block diagram is the basic concept of any cryptographic system.

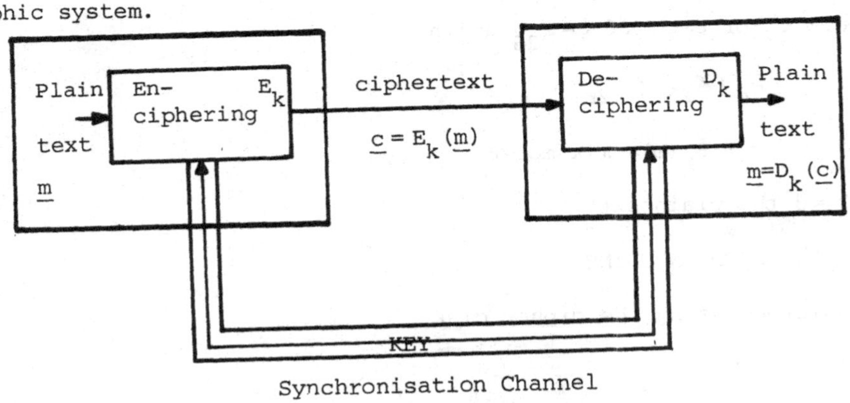

Picture 3.

The transmitter produces messages $\underline{m}$, the socalled plaintexts. These are sequences over some alphabet A . The set M of plaintexts is considered as a finite or countable cartesian power of the alphabet A , i.e. $M = \times A$. A Cipher System is a family $C = (E_k)_{k \in K}$ of ciphers $E_k : M \to M$ parametrized by the set K of keys. The keys are secretly agreed upon by the users. Then the message $\underline{m}$ is enciphered by E_k giving the ciphertext $\underline{c} = E_k(\underline{m})$. The process of deciphering is performed by applying the inverse map $D_k := E_k^{-1}$ to the ciphertext $\underline{c}$ thus giving

$D_k(\underline{c}) = D_k(E_k(\underline{m})) = E_k^{-1}E_k(\underline{m}) = \underline{m}$.

If the set M is a finite product, say $M = A^n$ the ciphers E_k are called <u>block ciphers</u> while in the other case, where $M = A \times A \times \ldots \times A \times \ldots$ consists of infinite sequences, the ciphers E_k are said to be <u>stream ciphers</u>. In some cases these definitions may turn out not to be distinctive.

The most famous example of an elementary cipher seems to that attributed to C.J. CAESAR.

The alphabet A,B,C,...,Z is canonically identified with the set of integers $\mathbb{Z}_{26}$mod26 . For each $k \in K = \mathbb{Z}_{26}$ define

$$E_k : \mathbb{Z}_{26} \to \mathbb{Z}_{26}$$

by $$m \to E_k(m) = m+k \bmod 26 .$$

Thus with $k = 3$ the plaintext

CISM IN UDINE

can be transformed using the cipher disk

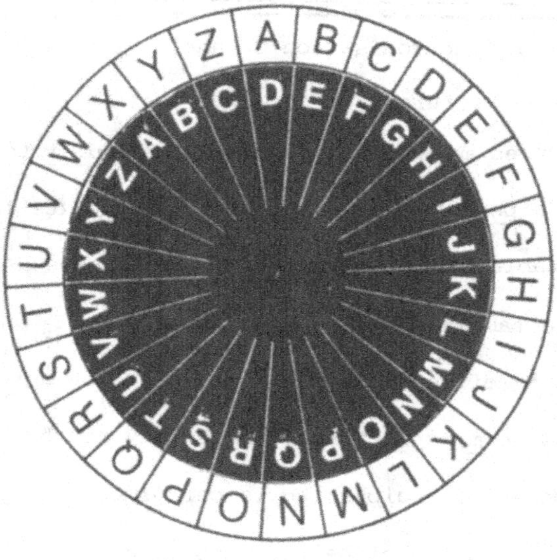

Picture 4: taken from: H.W. Franke, *Die geheime Nachricht*, Umschau/Ffm.

into the ciphertext

FMVP MQ XGMQH

It is unnecessary to say that such a cipher is easily broken (cf. Davio/ Goethals/Nemetz) by any interceptor. If an interceptor wants to cryptanalyse such a cipher system, he can perform a cryptanalysis based

- on a corresponding pair of plain- and ciphertext (given-plaintext-attack) or
- on large sets of ciphertext only.

The first kind of attack in case of a CAESAR is trivial. So is the second kind if the cryptanalist has reasonable a-priori-statistics of the underlying language at hand. All these socalled substitution ciphers, i.e. those where the symbols of the alphabet are permuted, are known to be extremely weak.

Among them are Thomas Jefferson's wheel cipher: Picture 5, below, cf. Kahn[1]

or the seemingly more complicated German ENIGMA

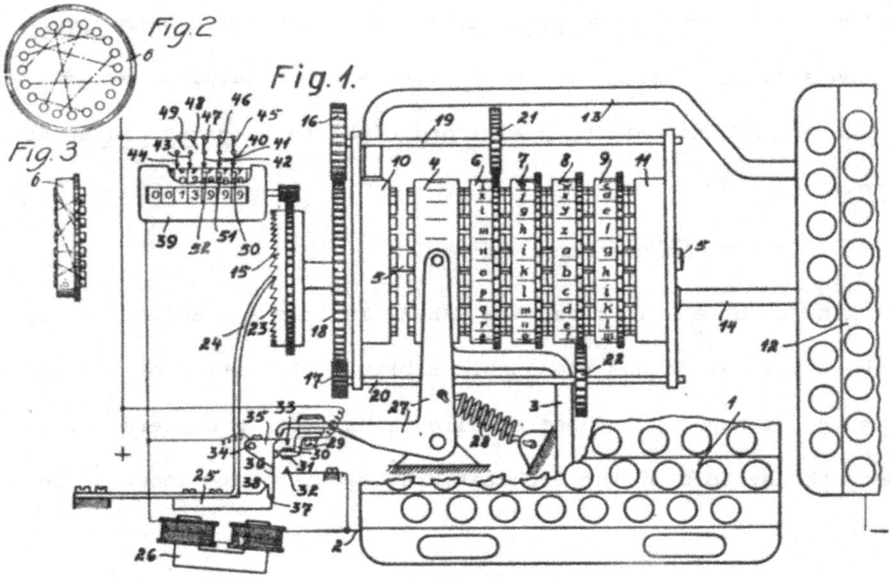

Arthur Scherbius' "Engima," U.S. Patent 1,657,411. Rotors 6, 7, 8, 9 are set to key NIAG. Figures 2 and 3 schematically show rotor connections

Picture 6: Functional drawing of an ENIGMA (taken from Kahn[1])

or the Hagelin Cryptograph C-48

(See picture 7 next page)

Picture 7.

All of them had been cryptanalysed a long time ago (cf. Kahn[1], Beker/Piper[2], Konheim[3]) sometimes even before their designers or users knew it. But with a slight modification these ciphers could have been secure.

3. A First look at Secure Stream Ciphers

This modification goes back to an idea due to G.S. Vernam, cf. Ryska/Herda[4]. Suppose that the plaintexts are already converted into a binary representation. Thus plaintexts are considered to be long (0,1)-sequences. The key-space M consists of the same set of sequences.

Mathematically speaking, the situation is

$$M = GF(2)^{\mathbb{N}}, \quad K = GF(2)^{\mathbb{N}}$$

and for each $\underline{k} \in K$ the ciphers $E_{\underline{k}} : M \to M$ are defined by

$$\underline{m} \to \underline{m} \oplus \underline{k}$$

where $\oplus$ means coordinatewise addition mod 2 .

Obviously each $E_{\underline{k}}$ is involutory, i.e. $E_{\underline{k}} = D_{\underline{k}}$, and the general concept of a Vernam cipher is this:

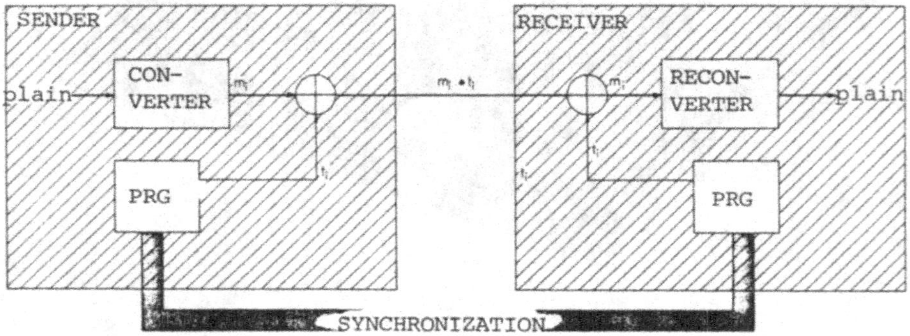

Picture 8: The Principle of a Vernam cipher

Now the problem is reduced to the study of the key generator:

If the key-generator is deterministic, it is likely to be insecure.

On the other hand, if the key generator is a random variable, the problems are these

1' : What is accepted to be random ?

(A key generator with a probability 10^{-100} of producing a '1' is hardly interesting)

2' : How to synchronize transmitter and receiver on the base of random experiment performed at both ends ?

The first famous solution to those problems seemed easy: take a binary version of some book in possession of both, use its contents as a key stream. The contribution of Nemetz shows you that this concept is cryptographically of no use, as it is insecure.

The second ingenious solution is to toss a coin, write the results down, store a copy, send a copy to the receiver and use this copy henceforth, as shown in the picture

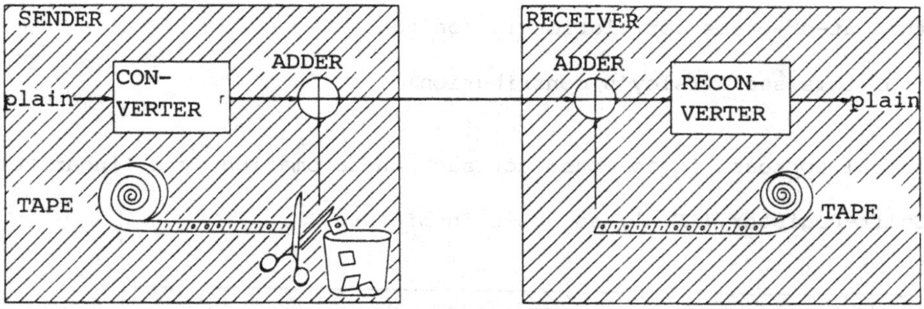

Picture 9: Vernam Cipher with One-time pad

This is the proper Vernam-cipher with a socalled on-time-pad.

It has the required features: for probabilistic reasons it is absolutely secure, as a

- ciphertext-only-attack is useless and a

- given-plaintext-attack is useless, as well.

But since everything has to be paid for, here is the major draw-back: such a system is expensive, slow, voluminous.

A third solution is to find a compromise: construct a key-generator which looks like a random-generator, but is cheaper and faster - the socalled pseudo-random-generator (PRG) .

In other words, a PRG is a deterministic automation which produces a 0,1-sequence that fulfils Golomb's randomness postulates, cf. Beker/Piper[3], of a random sequence, as there are

- Binomial-distribution with $p=\frac{1}{2}$, $q=\frac{1}{2}$
- the out-of-phase-correlation is constant (for this see Pursley's contribution).

A classical approach is to construct such an automation by a linear Feed-back-Shift-Register (LFSR), as shown in picture

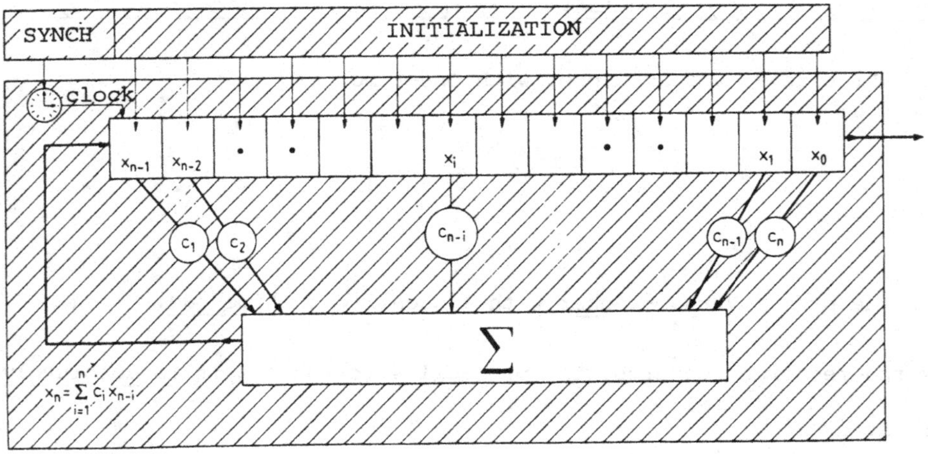

Picture 10: A linear Feedback Shift Register

4. An excursion to algebra

A linear feedback shift register produces a sequence

$$t = (t_o, t_1, \ldots, t_{n-1}, t_n, \ldots) \ .$$

This automation is started with the initial vector $\underline{t}^o = (t_o, \ldots, t_{n-1})$ which for cryptographic purposes has to be considered as the actual <u>primary key</u> . Once started, this automation runs autonomously. The outputs $t_n, t_{n+1}, \ldots$ are determined by the recursion, the socalled <u>feedback-equation</u>

$$(4.1) \qquad x_n = \sum_{i=1}^{n} c_i \, x_{n-i} \mod 2 \ .$$

In other words,

$$(4.2) \qquad t_{n+\tau} = \sum_{i=1}^{n} c_i \, t_{n+\tau-i} \mod 2 \text{ for all intergers } \tau \geqq 0 \ .$$

Translating this relation in matrix-form the equations (4.2) read

$$(t_\tau, t_{1+\tau}, \ldots, t_{n-1+\tau})A = (t_{1+\tau}, t_{2+\tau}, \ldots, t_{n+\tau}) \ ,$$

where the matrix A is given by

$$(4.3) \qquad A = \begin{pmatrix} 0 & \cdots & 0 & c_n \\ 1 & \ddots & & \vdots \\ 0 & \ddots & 0 & \vdots \\ \vdots & \ddots & \ddots & \vdots \\ 0 & \cdots & 0 \ 1 & c_1 \end{pmatrix}$$

From picture 10 it is clear that the assumption $c_n \neq 0$ is feasible. W.l.o.g. one then assumes $c_n = 1$.

Thus leads to the

Observaton 4.1: With the above assumption: A is nonsingular

The characteristic polynomial of A is

$$\chi_A(z) = z^n + \sum_{i=1}^{n} c_i \, z^{n-i} \in GF(2)[z]$$

As a linear feedback shift register is an autonomous finite state machine with an invertible transition function A , we conclude

Lemma 4.2: The sequence $\underline{t} = (t_i)_{i=o}^{\infty}$ is periodic.

Now let π be the shortest period of $\underline{t}$, i.e. the smallest positive integer p with

$$t_p = t_o, \; t_{p+1} = t_1, \ldots, t_{2p} = t_p \; .$$

Consider the "formal z-transform"

$$t(z) = \sum_{i=o}^{\infty} t_i z^i \; .$$

(Remark: t(z) is a formal power series, cf. Lüneburg[5]. Technically z^i can be considered as a timer)

Then the π-periodic behaviour can be used to describe t(z) as follows.

$$(4.4) \qquad t(z) = (t_o + t_1 z + \ldots t_{\pi-1} z^{\pi-1}) \cdot \sum_{j=o}^{\infty} z^{\pi j}$$

$$= \frac{t_o + t_1 z + \ldots + t_{\pi-1} z^{\pi-1}}{1 - z^{\pi}}$$

On the other hand t(z) can be described in terms of the recursion (4.2):

$$t(z) = (t_o+t_1z+\ldots+t_{n-1}z^{n-1}) + \sum_{\tau=o}^{\infty} t_{n+\tau}z^{n+\tau}$$

$$= (t_o+t_1z+\ldots+t_{n-1}z^{n-1}) + \sum_{\tau=o}^{\infty} \Big(\sum_{i=1}^{n} c_i t_{n+\ -i} \Big) z^{n+\tau}$$

$$= (t_o+t_1z+\ldots+t_{n-1}z^{n-1})$$

$$+ \sum_{i=1}^{n} \Big(\sum_{j=n-i}^{\infty} t_j z^j \Big) c_i z^i + \sum_{i=1}^{n} \Big(\sum_{j=o}^{n-i-1} t_j z^j \Big) c_i z^i$$

$$- \sum_{i=1}^{n} \Big(\sum_{j=o}^{n-i-1} t_j z^j \Big) c_i z^i$$

$$= (t_o+t_1z+\ldots+t_{n-1}z^{n-1})$$

$$+ \Big(\sum_{i=1}^{n} c_i z^i \Big) \cdot t(z)$$

$$- \sum_{i=n}^{n} \sum_{j=o}^{n-i-1} t_j \cdot c_i z^{i+j} \quad .$$

Denoting the last sum by p(z) it is immediate that t(z) reads

$$(4.6) \qquad t(z) = \Big(\sum_{i=1}^{n} c_i z^i \Big) t(z) + p(z)$$

where $\deg(p) \leqq n-1$.

With the socalled feed-back polynomial

$$q(z) := 1 + \sum_{i=1}^{n} c_i z^i = z^n \chi_A\Big(\frac{1}{z}\Big) \text{ in } GF(2)[z]$$

(the reciprocal polynomial of χ_A) we obtain

Theorem 4.3: $t(z) = \frac{p(z)}{p(z)} \in GF(2)[[z]]$.

Proof: Observe that

$$t(z)\Big(1+ \sum_{i=1}^{n} c_i z^i\Big) = p(z) \quad .$$

Combining the two results of (4.4) and Theorem 4.3, we obtain the important conclusion:

Corollary 4.4: $t(z) = \frac{p(z)}{q(z)} = \frac{t_0+t_1z+\ldots t_{n-1}z^{n-1}}{1-2^{\pi}}$

Corollary 4.5: If $q(z) \in GF(2)[z]$ is irreducible, then $q(z)$ divides $1-2^{\pi}$.

In other words, $q(z)$ is a "cyclotomic" polynomial for the period π .

Going back to the application of designing a one-time-pad cipher system we need to construct a LFSR with an extremely long period π .

For this purpose let $q(z) \in GF(2)[z]$ be a socalled primitive polynomial of $\deg q = n$, i.e. the roots of $q(z)$ have multiplicative order 2^n-1 . Then the corresponding LFSR with feedback polynomial q has the period $\pi = 2^n-1$. It is relatively easy to show that this LFSR produces a pseudo-random sequence fulfilling the Golomb-postulates for a Pseudo-noise-sequence.

Conclusion 4.6: A linear feedback shift register with primitve feedback-polynomial produces a PN-sequence. Such a device is easily implemented (by Standard IC's), cheap and fast.

But: It is totally insecure from a cryptographic point of view.

5. How to break a stream cipher which is generated by a LFSR-sequence:

For this we recall that the sequence $\underline{t} = (t_0, t_1, \ldots, t_{n-1}, t_n, \ldots)$ can be described via the transition matrix

$$A = \begin{pmatrix} 0 & \cdots & \cdots & \cdots & c_n \\ 1 & \ddots & & & \vdots \\ 0 & \ddots & \ddots & & \vdots \\ \vdots & \ddots & \ddots & 0 & \vdots \\ 0 & \cdots & 0 & 1 & c_1 \end{pmatrix} .$$

For any time r O define the internal state vector (cf. Pict. 10)

$$\underline{t}^{\tau} = (t_{\tau},\dots,t_{n-1+\tau}) \;.$$

then

$$\underline{t}^{\tau} = \underline{t}^{o}\, A^{\tau} \qquad .$$

Since A has companion-matrix form, the characteristic polynomial χ_A and the minimal polynomial μ_A of A coincide, cf. Lüneburg[5]. We thus have

Lemma 5.1: Assume χ_A is irreducible. Then for all $i \geq 0$ the $(n \times n)$-matrix

$$T^{(i)} = \begin{pmatrix} \underline{t}^{i} \\ \underline{t}^{i+1} \\ \vdots \\ \underline{t}^{i+n-1} \end{pmatrix}$$

formed by the n consecutive vectors $\underline{t}^{i},\dots,\underline{t}^{i+n-1}$ is nonsingular and for all $i \geq 0$

$$T^{(i)} A = T^{(i+1)} \qquad .$$

Corollary 5.2: The sequence $\underline{t} = (t_o,\dots,t_{n-1},t_n,\dots)$ is completely determined by at most 2n consecutive entries.

Conclusion 5.3: Under a given-plaintext attack, an analyst only needs 2 n bits to break the cipher, although the period is very long, namely 2^n-1 .

Thus we draw the main

Conclusion 5.4: A PN-sequence has to be considered under both

- probabilistic aspects and the
- aspect of complexity theory

6. An Analysis of Binary Deterministic Finite State Machines:

A binary deterministic finite state machine, a socalled Boolean automaton, is best described as follows:

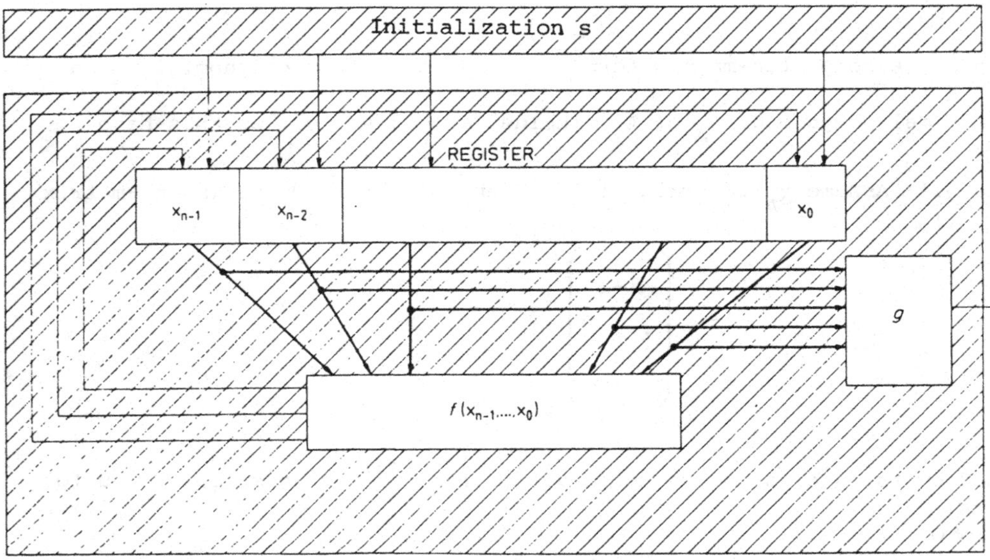

Picture 11: A Boolean Automaton

If it runs autonomously, it eventually becomes periodic with period π .

As before

$$(6.1) \qquad t(z) = a(z) + \frac{r(z)}{1-z^{\pi}}$$

where a(z) is the polynomial describing the pre-period, which is due to the behavior of general Boolean automata, cf. Jennings[6], Beth/Heß/Wirl[11] .

Henceforth we only shall consider the periodic part $\frac{r(z)}{1-2^{\pi}}$ which after cancelling common factors has the form

$$\frac{r(z)}{1-z^{\pi}} = \frac{P(z)}{Q(z)} \quad \text{where}$$

$P, Q \in GF(2)[z]$ are coprime. Therefore the periodic part could as well have been produced by a LFSR with feedback polynomial $Q(z)$.

Definition 6.1: As a measure of complexity of the Boolean automation define the linear equivalent $L = \deg(Q(z))$.

Question 6.2: How would a cryptanalyst proceed with a non-linear Boolean automaton ?

We shall see now that this problem is closely related to the problem of decoding cyclic codes, as described in the contribution of Blahut.

The analyst assumes the existence of an equation $t(z) = \frac{P(z)}{Q(z)}$ or , equivalently, $t(z) \cdot Q(z) = P(z)$.

At the time instance $\tau = k$ he only knows the series $t(z)$ up to its k-th term, i.e. he considers

$$t(z) \bmod z^{k+1} .$$

Recursively he starts a series of interpolating polynomials $Q_k(z)$, $P_k(z)$ fulfilling $t(z) \cdot Q_k(z) \equiv P_k(z) \bmod z^{k+1}$ at each time $k \geq 0$ as follows:

Algorithm 6.3 (Massey-Berlekamp): cf. Berlekamp[7]

Start for $\tau = 0$

Suppose that until $\tau = k$ there are $Q_k(z)$ and $P_k(z)$ such that

$$t(z) \cdot Q_k(z) \equiv P_k(z) \bmod z^{k+1}$$

Then $\tau := k+1$.

If $t(z) \cdot Q_k(z) \equiv P_k(z) \bmod z^{k+2}$

define $Q_{k+1}(z) := Q_k(z)$

and $P_{k+1}(z) := P_k(z)$,

else, for $t(z) \cdot Q_k(z) \equiv P_k(z) + F_{k+1}, z^{k+1} \bmod z^{k+2}$,

define $Q_{k+1}(z) := Q_k(z) - F_{k+1} \cdot z \cdot G_k(z)$

and $P_{k+1}(z) = P_k(z) - F_{k+1} \cdot z \cdot (t_k(z)$

where G_k , $H_k \in GF(2)[z]$ are of minimal degree such that

$$t(z) \cdot G_k(z) \equiv H_k(z) + z^k \bmod z^{k+1} \quad .$$

Lemma 6.4: If the linear equivalent L of t(z) is L = N , then this algorithm succeeds in O(2N) computational steps in determining n , P(z) and Q(z) .

Corollary 6.5: After the events $F_{L+1} = \ldots F_{L+r} = 0$ the decision that $t(z) = \frac{P_L(z)}{Q_L(z)}$ fails with probability $\frac{1}{2^r}$.

Conclusion 6.6: Never use Pseudo-Random generators with a small linear equivalent !

7. How to construct "good" P-R-generators ?

The idea of designing acceptable P-R-generators by composing them from LFSR's has occured in many sources, eg., cf. Ryska/Herda[4] , Beker/Piper[3].

For this let $LFSR_i$ (i=1,2) be two linear generators with feedback polynomial $q_i(z)$, $\deg(q_i)=n_i$, where the q_i are coprime and irreducible.

Design a new generator as follows.

Example 7.1: $f(x,y) = x+y \bmod 2$

What is the linear equivalent of the sum sequence $s_t = u_t + v_t$?

From section 4 we know, that there are polynomials $p_i(z)$, $\deg(p_i) \le n_i - 1$ such that

$$u(z) = \sum_{t=o}^{\infty} u_t z^t = \frac{p_1(z)}{q_1(z)}$$

$$v(z) = \sum_{t=o}^{\infty} v_t z^t = \frac{p_2(z)}{q_2(z)}$$

Thus $(u+v)(z) = u(z) + v(z) \bmod 2 = \dfrac{p_1(z)q_2(z) + p_2(z)q_1(z)}{q_1(z)q_2(z)}$.

As q_1 and q_2 are irreducible and coprime this is the rational function of minimal degree giving $u(z) + u(z)$.

Thus the linear equivalent $L(u+v) = L(u) + L(v) = n_1 + n_2$.

Therefore this approach is not likely to give any more complicated sequence.

Problem 7.2: If $f(x,y) = x \cdot y \bmod 2$.

What is the l.e. of the sequence produced by the following composition ?

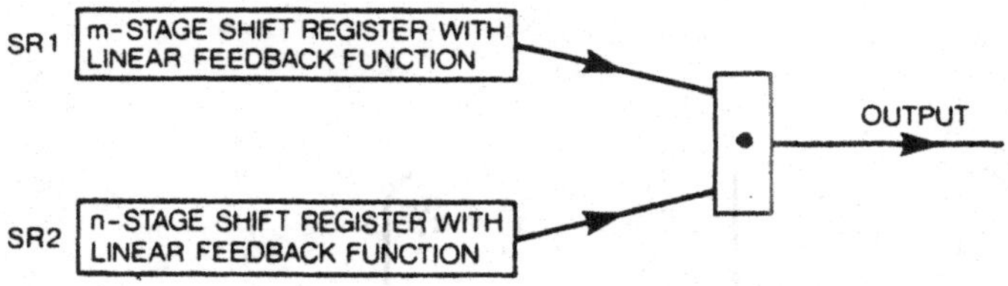

Picture 12: The AND-gate Generator

Here again a short look at the tools of linear algebra will be helpful:

For LFSR with feedback polynomial $q(z)$ the output sequence $(s_t)_t$ is described via

$$(7.1) \qquad s(z) = \sum_{t=0}^{\infty} s_t z^t = \frac{p(z)}{q(z)}$$

$$(7.2) \qquad (s_\tau, \ldots, s_{n-1+\tau}) = (s_0, \ldots, s_{n-1}) \cdot A^\tau$$

$$\text{where } A = \begin{pmatrix} 0 & \cdots & \cdots & \cdots & c_n \\ 1 & & & & \vdots \\ 0 & & & & \vdots \\ \vdots & & & & \vdots \\ \vdots & & & 0 & \vdots \\ 0 & \cdots & 0 & 1 & c_1 \end{pmatrix} \quad \text{and } q(z) = z^n \chi_A(\tfrac{1}{z}) \quad .$$

The reciprocal eigenvalues $\alpha_1, \ldots, \alpha_n$ of A are the zeros of $q(z)$. They are all distinct, as q is supposed to be irreducible.

Then the <u>Fourier matrix</u>

$$F = \begin{pmatrix} \alpha_1^0 & \alpha_1^1 & \cdots & \alpha_1^{n-1} \\ \vdots & & & \vdots \\ \alpha_n^0 & \alpha_n^1 & \cdots & \alpha_n^{n-1} \end{pmatrix}$$

is non-singular. Thus there are uniquely determined elements $\delta_1, \ldots, \delta_n \in GF(2^n)$ such that $(s_0, \ldots, s_{n-1}) = (\delta_1, \ldots, \delta_n) \cdot F$. Cf. Selmer[8].

Observing

$$\left(\begin{matrix} \alpha_1^1 & \cdots & \alpha_1^{n-1} \\ \vdots & & \vdots \\ \alpha_n^1 & \cdots & \alpha_n^{n-1} \end{matrix} \;\middle|\; \begin{matrix} \sum_{i=0}^{n-1} c_{n-i} \alpha_1^i = \alpha_1^n \\ \vdots \\ \sum_{i=0}^{n-1} c_{n-i} \alpha_n^i = \alpha_n^n \end{matrix} \right)$$

we conclude for all integers t

$$s_t = \sum_{i=1}^{n} \delta_i \alpha_i^t \quad .$$

The reasoning shows that if $q(z)$ is irreducible over GF(2) the δ_i $(i=1,\ldots,n)$ are conjugate, i.e. $\delta_i = \delta^{2^i}$ for some $\delta \in GF(2^n)$.

Applying this to problem 7.2 we obtain a nice expression for the sequence $(u_t \cdot v_t)_t$.

We know, that for each positive integer t

$$u_t = \sum_{i=1}^{n_1} \delta_i \alpha_i^t \quad \text{(where the } \alpha_i \text{ are the zeros of } q_1)$$

$$v_t = \sum_{j=1}^{n_2} \gamma_j \beta_j^t \quad \text{(where the } \beta_j \text{ are the zeros of } q_2) .$$

Thus

$$u_t \cdot v_t = \sum_{j=1}^{n_1} \sum_{i=1}^{n_2} \delta_i \gamma_j (\alpha_i \beta_j)^t .$$

If $(n_1, n_2) = 1$ then the smallest field containing all elements $\alpha_i \beta_j$ is $GF(2^{n_1 \cdot n_2})$ and the minimal polynomial of $u_t \cdot v_t$ is

$$q_1 \wedge q_2(z) = \prod_{i,j} (z - \alpha_i \beta_j) .$$

Thus we have

Lemma 7.3: The linear equivalent of the AND-gate-generator (Jeffe, cf. Ryska/Herda[4]) is $L = n_1 \cdot n_2$.

Conclusion 7.4: The AND-gate generator can be used to gain considerably large complexity - at the expense of probabilistic properties: the probability of producing a "0" is about 3/4 . This violates Golomb's postulates.

A better construction, dealing with both aspects, has been developed and investigated by Beker/Piper[2], Jennings[6], Beth/Hain[9] in the years 1979-1981

The device is called a multiplexer-generator and works according to the following block diagram

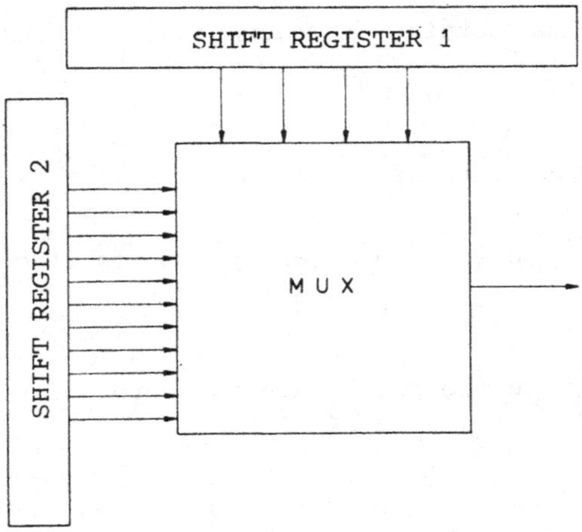

Picture 13: The MUX-Generator

Assume that the feedback-polynomials $q_i(z)$ of the $LFSR_i$ are coprime and primitive of degree $\deg(q_i) = n_i$, $(n_1, n_2) = 1$.

Conclusion 7.5: Then for each k , $1 \leq k \leq 2^{n_1}$ and $2^k \leq n_2$ is a MUX-Generator with

- linear equivalent $L = n_2 \cdot \sum_{i=0}^{k} \binom{n_1}{i}$ and
- period $\Pi = (2^{n_1}-1)(2^{n_2}-q)$ and
- feasible probabilistic behaviour as requested.

For a detailed study, cf. Beker/Piper[2], Jennings[6], Hain[10].

Another example is the socalled switch-on-generator (Beth/Hain[9]).

Take r $LFSR_i$, (i=1,...,r) which are primitive of coprime degrees n_i .

The block diagram describes its functions.

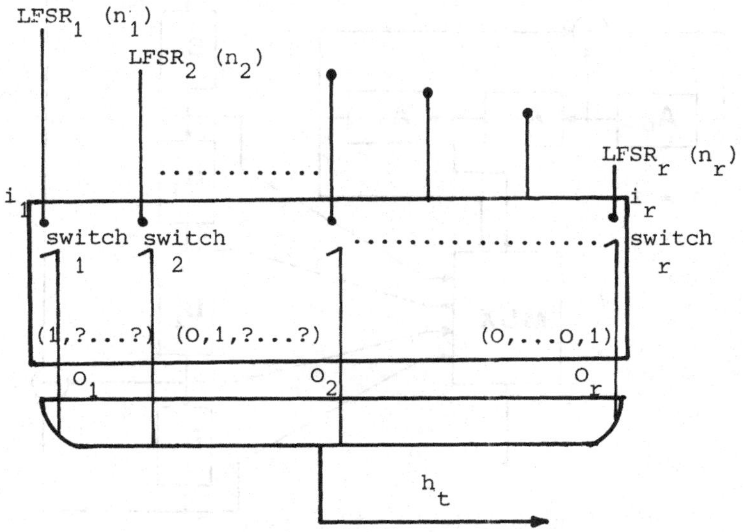

Picture 14: A Switch-On Generator

If the jth switch is on, then output O_j produces a "1" iff the input pattern $(i_1,\ldots,i_r)$ equals $(O_1,\ldots,O_{j-1},1_j,?,?,?,\ldots?)$ where ? denotes "don't care" .

The sequence h_t produced can be adjusted to any wanted probability of "1"

$$0 < \text{Prob}\ (1) = \sum_{i=1}^{r} \frac{\varepsilon_i}{2^i} < 1 \text{ with}$$

$$\varepsilon_i = \begin{cases} 0 & \text{iff } O_i \text{ is "OFF"} \\ 1 & \text{iff } O_i \text{ is "ON"} \end{cases} .$$

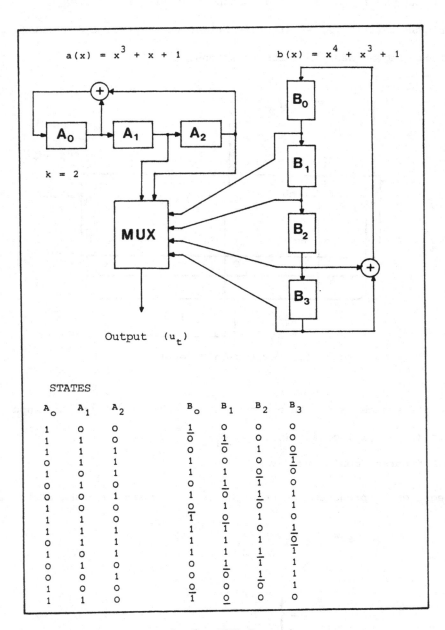

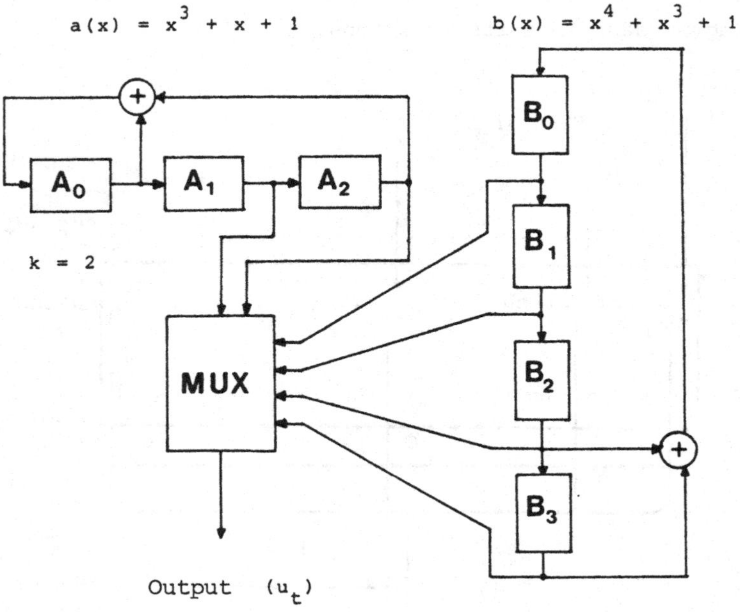

STATES

A_o	A_1	A_2	B_o	B_1	B_2	B_3
1	0	0	1	0	0	0
1	1	0	0	1	0	0
1	1	1	0	0	1	0
0	1	1	1	0	0	1
1	0	1	1	1	0	0
0	1	0	0	1	1	0
0	0	1	1	0	1	1
1	0	0	0	1	0	1
1	1	0	1	0	1	0
1	1	1	1	1	0	1
0	1	1	1	1	1	0
1	0	1	1	1	1	1
0	1	0	0	1	1	1
0	0	1	0	0	1	1
1	0	0	0	0	0	1
1	1	0	1	0	0	0

Picture 15: An example for MUX-Generator

Conclusion 7.6: The linear equivalent then is $L = \prod_{i=1}^{r} n_i^{\varepsilon_i}$

and the period equals $\pi = \prod_{i=1}^{r} (2^{n_i}-1)^{\varepsilon_i}$.

This device has successfully been employed in several transmission channels, cf. Beth/Hain et.al.[9].

Finally an important cryptographic device, that Data Encryption Standard (DES) of the NBS has to be mentioned. Its operation in cipher feedback mode, cf. Konheim[3], Ryska/Herda[4] seems to deliver a rather strong stream cipher. Unfortunately, no approach on determining its linear equivalent seems to be known.

8. References:

1. Kahn, *The Codebreakers*, McMillan, 1982

2. Beker/Piper, *Cipher-System - The Protection of Communications*, Northwood, 1982

3. Konheim, *Cryptography: A Primer*, Wiley, 1981

4. Ryska/Herda, *Kryptographische Verfahren in der Datenverarbeitung*, Springer, 1980

5. Lüneburg, *Galoisfelder, Kreisteilungskörper und Schieberegisterfolgen*, Bibliographisches Institut, 1979

6. Jennings, *PHD-Thesis*, University of London, 1980

7. Berlekamp, *Algebraic Coding Theory*, Mac-Graw-Hill, 1968

8. Selmer, *Linear Recurrences over Finite Fields*, Technical Report, University of Bergen, 1968

9. Beth/Hain et.al., *Materialien zur Codierungstheorie II*, University of Erlangen, 1979

10. Hain, *Diplomarbeit*, University of Erlangen, 1982

11. Beth/Heß/Wirl, *Kryptographie - Eine Einführung*, Teubner, Stuttgart

SECRET SHARING SYSTEMS

by S. Harari

I. Introduction

All through this paper the notion of secret that we will discuss is a notion linked to the use of a digital quantity called "key" for a given equipment that is turned on by this key. To be more precise: the value of a variable is secret for a given piece of equipment if the possibility of use of this equipment, by trying all possible values of the variable is very long compared to a given duration.

Example:

for a piece of equipment whose response delay is one hour, a variable is secret if it has 2^{20} possibilities. I.e. the variable is secret if its entropy is at least 20 bits.

If the response delay is one millisecond, a variable will be secret if it has 2^{60} possible values, i.e. its entropy is at least 60 bits. In the rest of the paper, a binary quantity will be said to be "secret" if it is 60 bits long.

This notion of a secret quantity, although not very classical, is of great use in the case of financial terminals, or more generally, to solve the problem of access to remote computer equipment (terminals, modems, etc.)

II. The problem of sharing a secret quantity

Let S be a secret quantity in the above sense. The problem we will try to solve is the following:

how can we divide the information bits of S between a partners such a way that any b partners can find S from their share of secret, and that any (b − 1) cannot. We also require b to be strictly less than a.

The reason for this last condition is that if b was equal to a the system would be "hostage" of each of the secret holders. In that case, the absence of only one of the secret-holders, would stop the others from finding S from the quantity they legitimately own.

III. Information theory of secret sharing systems

Clearly, just dividing the bits of which S are made between the intended secret sharers, will not solve the problem. Some form of coding is needed. We have the following diagram.

Fig. 1 : secret sharing system

The data D to be shared, is to be coded. The resulting codeword is S. The bits of S are shared between the a legitimate secret sharers.

At the decoding end, the decoder collects the data from at least b secret sharers. From this data it finds the quantity S. The basic idea is to give each secret sharer a quantity of information, whose entropy H satisfies the following conditions.

1) $(a - b) \cdot H$; is an entropy that can be recovered by some algorithm.

2) $(a - b + 1) \cdot H$; cannot be recovered; by any algorithm in the given amount of time.

From these general considerations one can derive bounds on the size of a discrete code, designed to solve this problem.

III.2 Bounds on the size of the code

Suppose some kind of block code is used to share the secret; with information symbols from an alphabet of size q.

The entropy of each symbol is $s = 1/q \log_2 q$.

Let n be the number of symbols given to each secret holder (we can suppose that each secret holders has the same number of symbols).

We have the following condition, in order that condition (*) be met.

$$n \cdot s \geqslant 60 .$$

This implies that the length N of S satisfies the following

$$N = bn \geqslant b \cdot \frac{60}{s}$$

III.3 The use of error correcting codes

Block error correcting codes satisfy the conditions developed in (3.1.1.).

Suppose we consider an error correcting code of length N and minimum distance D; over an alphabet with q symbols. Let t = [d – 1/2] be the number of symbol errors the code can correct.

We can interpret this by the fact that the code with its error correction algorithm can lift the uncertainty on an a word if the entropy is less than t · s; and that the code cannot (*) lift the uncertainty if the entropy of the word under consideration is more than t · s.

One can easily find codes with parameters (N, D, q) satisfying these conditions.

IV An example of implementation

Let us consider an error correcting Reed Solomon on F_q of length N = q – 1. The entropy of each symbol is s = – log q/q.

Therefore if the code is used to share a secret between a secret sharers we must have

$$N \cdot s \geqslant a \cdot 60$$

The Reed Solomon over F_{42} has length 46. Each symbol has an entropy of the order of 5 bits.

The entropy of a codeword can at most be

The code can be used to hsare a secret between at most

$$\frac{47 \times 5}{60} \cong 4 \text{ sharers}$$

Each share of secret is made of 11 symbols, together with their position in the codeword. Taking a D = 24 code we have a system of type (4,3). We cannot go any further in reducing the number of necessary sharers by using this code, and only one codeword.

IV.2 Decoding

The procedure to obtain the secret quantity S is as follows: all the present secret sharers feed their shares of the secret in the correct position of the word to be decoded. The missing symbols are arbitrarily set to 0. The decoding procedure of the code is then set into action.

If instead we had used the Reed Solomon of length 126 on F_{127} , we could have settled a secret between

(*) Note: This supposes that there are more than 2^{60} codewords at distance d from the codeword under consideration.

$$\frac{126 \times 7}{60} \cong 14 \text{ sharers .}$$

Each share of secret is made of 9 symbols.

Taking a $D = 18$ code we have a system of type (14, 13)

$D = 36$ yields a system of type (14, 12)

$D = 54$ yields a system of type (14, 11).

IV.3 Other type of parameters

By not giving away all possible symbols one can modify the secret sharing capacity.

A system of type (a,b) will yield a system of type $(a - i, b - i)$ for $1 \leqslant i \leqslant b - 1$.

One can also easily see that a system of type (a,b) is a system of type (a,b′) with $b \leqslant b'$; if some precautions are taken.

IV.4 Some general binary and non binary error correcting codes

One can see from the above examples, that the codes that are needed, have to have a very large minimum distance. This usually implies that the codes that are used be non binary. We give a table of some codes, the maximum number of secret sharers, and the minimum number of sharers that are required to be present in order to be able to get the secret quantity back.

The fact that the minimum distance has to be large for binary codes, implies that we use very long codes, with very high minimum distance. This in turn implies that the decoding procedure will be very long and costly in software.

IV.5 Interleaving

One can overcome this difficulty by using a shorter code, with many codewords.

The secret S will be a set of codewords; the sharers having each a few bits from every codeword. It is essential that no one sharer have all the bits from one codeword.

The decoding procedure applied to every codeword, allows then to obtain S from the available data.

V. Conclusion

Codes designed to share secrets have to be, a priori, block codes. Available error correcting codes, meet the requirements of secret sharing if used in an appropriate manner. Although we have developed an information theory of these codes it remains to be proved that error correcting codes are an optimal solution for such a problem. However the very good knowledge we have of error correcting codes, and their decoding algorithms, makes it very hard for them to be surpassed in practical use.

REFERENCE

F.J. MacWilliams, N.J.A. Sloane: The Theory of Error Correcting Codes, North Holland Publishing Company 1977.

Table of possible systems based upon Reed Solomon codes:

Length of the code	# symbols per sharer	a max.	b min.
47	9	4	2
101	10	10	4
149	10	18	11
197	8	24	15
293	8	35	19

KEY MANAGEMENT IN DATA BANKS

by S. Harari

Summary

In this conference we consider the problem of key manaqement in a data bank with enciphered files. We do not consider the problem of encruption of data links leading to the data bank. We assume a protection hierarchy exists and is in charge of the problem. We consider the problem of the overhead needed to manage the different keys, and the constraints that reconfiguration of the data bank puts on the system. We show that a "natural" solution to this problem is not optimal. We give another solution; formally more restrictive, but practically meeting all the constraints of our problem.

Introduction

We consider a data bank made of files that are enciphered. We suppose that the algorithms used to encipher and decipher these files need two secret (maybe one) keys for their implementation. We will be interested in managing those keys, by that we mean, that we would like to establish automatic procedures of access, that do not need a big amount exchange of keys.

The aim of this cryptography is, for example, to eliminate people that have not subscribed to the services of the data bank.

We suppose that the data links between the user and the data bank are safe, i.e. are enciphered, and the key management for this type of cryptography is done by a hierarchy of secret, which is independent of the data bank.

1. The natural solution

Let F be a file enciphered by a key K_F and deciphered by a key K_F'. For a subscriber A to the data bank, in order to be able to have access to the file F, it is necessary for the file F; to know that A is allowed to have access to it. This step is to eliminate illegitimate users. A also must possess the key K_F.

One can immediately do the following remark.

– The key K_F is the same for everybody. The discrimination between regular subscribers, and illegitimate users is done at the level of the file F. Because of the number of files, this checking must be left quite light.
– Such a system is quite vulnerable to substitution. For a given file F. The key is given to many users; who are not very keen on keeping it safe. The theft of the key K_F, does not compromise their subscription to the data bank. The illegitimate user will on the other hand only have to act as a regular subscriber to be able to have access to the file F, with the key K_F he has illegally obtained.
– The necessary periodic change of the keys K_F is a very difficult operation. For a given file the new key is to be given to all the subscribers having access to the file F. This must be done for every file F. This is a difficult, operation because of the volume of the data involved. If a bank has 10.000 subscribers and 1000 files per subscriber, this implies that 10^7 keys have to be exchanged periodically.

In use, this system has many shortcomings.

a) For a given file F, the technical overhead is very heady (10.000 names of subscribers).
b) For a given subscriber, the key management is a difficult problem. The risk of theft is quite high, the risk of loss also, and the risk of using the wrong key can hinder operations considerably. (a subscriber can have 1000 keys. For this size, a mislabelling of the keys can be as damageable as the loss of the keys.)
c) For the management of the data bank, the reconfiguration of the subscribers can only be done when the keys are damaged. The loss of the key of one file by one subscriber, necessitates that all subscribers to that file have their key for the file changed. This can increase considerably the amount of keys being exchanged. d.

– When a key change is in progress for a file F, the file F is not accessible. This means that globally the data bank will not be available for a large percentage of time.
– No selective access to a file is possible. Everybody is treated as equal by a file.

2. The intermediary file solution

We are now going to describe a procedure, with an intermediary file, encrypted, between the user and the data bank file; called thereafter a working file. The description of the set up is the following:

Let F_1 , F_2 , . . . be encrypted working files, decipherable by the keys K_{F_1} , K_{F_2}, . . .

A subscriber A is in possession of one key K_A ; giving him access to an intermediary file I_A. The file I_A has stored the cryptograms K_{F_1}, K_{F_2}, . . . of the keys of the files K_{F_1}, K_{F_2}, . . . to which A has legitimate access.

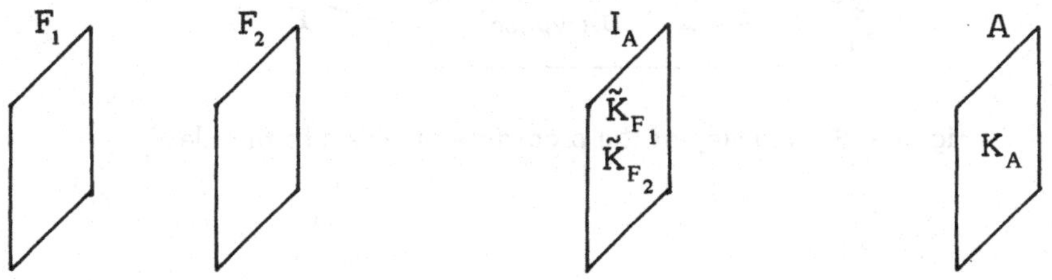

Fig. 1 - Set up of the procedure using an intermediary file.

In order for subscriber A to have access to a working file F_1 , he must first use his key K_A, to obtain K_{F_1} from K_{F_1} in I_A ; by using the decryption algorithm.

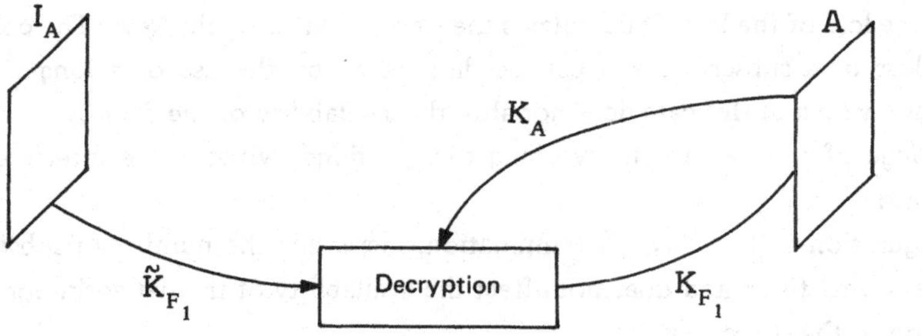

Fig. 2 - First step of the procedure of access to the file F_1 .

The second step of the procedure is to obtain the file F_1 , by using the decryption machine, with the key K_{F_1} , and as cryptogram the content of the enciphered version of F_1 .

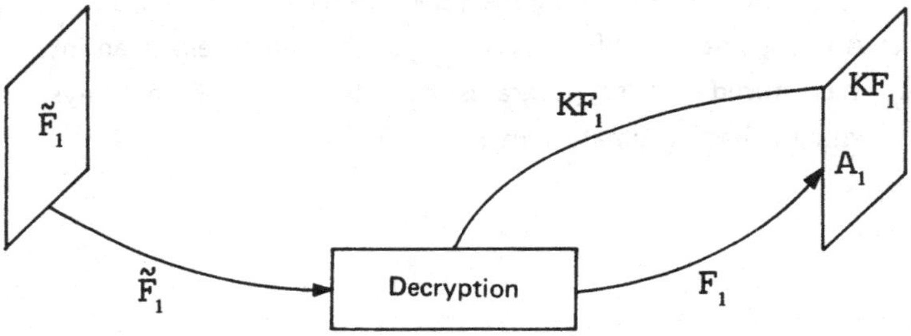

Fig. 3 - Second step of the procedure of access to the file F_1 .

There is need for a randomizing procedure in obtaining the key K_{F_1}, so that a given key is valid for one use, i.e. another attempt to obtain the file F_1 with the same key K_{F_1} , must fail. This is done in order for the procedure to be the same at all times.

This procedure is longer than the preceeding one. In order to have access to file F_1 , two decryption operations must be done, whereas one only was needed in the preceeding one.

The advantages of this type of procedure are the following.

1) The overhead for each file F is limited to one key, whatever is the number of subscribers having access to F.
2) Each subscriber to the data bank has only one key in his possession. This minimizes the risk of the loss of the key. It eliminates the risk of confusing the keys. The consequences of the loss of a subscriber key can be diminished by the use of a long "handshake" procedure, which in this case does not alter the availability of the data bank.
3) The change of the key to the working files, is done, without the interference of the subscribers.
4) Reconfiguration of the network (elimination, increasing the number of subscribers) can be done in real time, and does not affect the availability of the network, nor the data in possession of the subscribers.
5) Hierarchy is made possible, by giving a tree structure to the intermediary files I_A .
6) Change of subscribers keys, involves a minimum amount of key traffic, and can be done at different frequencies for different subscribers, allowing for different "safety ratings" for different subscribers.

3. Conclusion

We have presented two procedures for key management in a data bank. The advantages of the second over the first are numerous. They are obtained at the cost of a supplementary decryption procedure. It is a kind of memory, computation tradeoff. This tradeoff is worthwhile in many cases, since a network, or data bank necessitates a large amount of auxiliary computing power.

REFERENCE

Gudes E., Key Management . . ., IEEE Transactions on software engineering, vol. SE n. 5, Sept. 1981.

ELECTRONIC SIGNATURE FUNCTIONS

S. Harari

In this first part we examine all the conditions that must be satisfied by the electronic signature, and examine the mathematical consequences.

All the quantities that will be dealt with in the sequel are digital quantities, binary or decimal. The entropy of a number N is its length if it is expressed in bits, or $\log_2(N)$ otherwise.

A given number N will be a safe number if its entropy is at least 60 bits.

a) The signature S of a message M by a person P is a safe quantity, that belongs to P. Ideally it is a safe quantity that P determines, and that only P can determine. This implies the existence of a secret function; and that the output of the function is a safe quantity.

b) Checking the signature S must be easy.
This implies that there is a function whose output is the signature is valid, or invalid. This function must be computable by anybody. Therefore it must not contain any secret data. The knowledge of that function must not give any indication on the signature function. It must be easily computable. In particular it should not necessitate big memory space, or big delays.

c) The signature function must depend on the message M to be signed; in a random but computable way.
If that was not the case, the signature of a lessage M, would then be valid as the signature of a message M'. This is true even for the message M: The signature of a given message M should not be twice the same. This implies that the signature function should have another variable.

The signature function s, and the checking function v are therefore as follows.

$$\mathcal{P}_x \mathcal{M}_x \mathcal{R} \xrightarrow{s} \mathcal{S} \underset{v}{\rightarrow} \mathcal{P}_x \mathcal{M} .$$

P is the set of possible persons, m the set of possible messages and R a set of random numbers. The function s must be kept secret, that is the legitimate owner should not disclose it. The secret in fact can reside in some parameters, the general structure of s can be known, as we shall see later.

The function v is a partial inverse to s, that is for a message M, a person P, a random quantity R we must have

$$v(P, M, R) = (P, M) .$$

II. An example of signature function; Rabbin's algorithm

Let p and q be two very larqe prime and n = p q. Rabbin [1] has shown that finding a solution to the equation $x^2 = a \bmod n$ is equivalent to the prime decomposition of n.

The algorithm is as follows. The legitimate owner of the integers p and q solves the equation $x^2 = M \bmod n$; as we shall see in the sequel.

The signature of M is (M,x). Checking the signature is just checking the equation $x^2 = M \bmod n$, which can be done even if one does not know the prime decomposition of n.

We shall now show how to solve the equation $x^2 = a \bmod p$ where p is a prime. Doing the same thing mod q allows us to solve the equation mod n by using the chinese remainder theorem.

Two cases arise, when trying to solve; the equation $x^2 = a \bmod p$, depending on the fact that $p = 4k - 1$, or $p = 4k + 1$.

a) Case where p = 4k1

In any finite field half of the nonzero elements are quadratic residues, and half are not. Let a be a quadratic residue. One can easily check that $a^{1/2} = a^{p+1/2} \bmod p$.

The solution of the equation $x^2 = a \bmod p$ is done by an exponentiation operation for which there exists very efficient algorithms.

b) Case where p = 4k + 1

The algorithm m in this case, a little less straightforward necessitates extra definitions:

In a finite field of order p, two elements will be said to be of the same type if they are both either, quadratic residues, or non residues.

In such a field the quadratic residues are all the roots of the equations

$$x^{(p-1)/2} - 1 = 0 \mod p.$$

Let f(x) be a polynomial of degree 2; and let

$(f(x),\ x^{p-1/2} - 1)$ be the G.C.D. of these two polynomials.

$(f(x),\ x^{p-1/2} - 1)$ be the G.C.D. of these two polynomials.

The following cases can occur

$$(x^{p-1/2} - 1,\ f(x)) \begin{cases} = 1 \\ = (x-\alpha) \\ = f(x)\,. \end{cases}$$

a) the G.C.D. of the two polynomials is 1 if none of the zeros of $x^{p-1/2} - 1$ is a quadratic residue.

b) the G.C.D. is a polynomial of degree 1 if one of the two zeros is a quadratic residue. In this case one obtains the other zero of f(x) just by dividing f(x) by $x\ \alpha$.

c) the G.C.D. is f(x). This case happens if both the zeros are quadratic residues.

Clearly case b) is the favourable one. To be able to get back to case b, let us consider the following q(x).

$q(x) = x - \alpha / x - \beta$, where α and β are the unknown roots of f(x). Now consider the mapping φ from $F_q - \{-\beta\}$ into F_q :

$$\varphi : \gamma \rightarrow q(x-\gamma) = \frac{x-\gamma-\alpha}{x-\gamma-\beta}$$

$\varphi(\gamma)$ is a quadratic residue if $\alpha + \gamma$ and $\beta + \gamma$ are of the same type, $\varphi(\gamma)$ is a non residue if $\alpha + \gamma$ and $\beta + \gamma$ are not of the same type.

The mapping $F_q - \{-\beta\} \rightarrow F_q$, $\gamma \rightarrow \varphi(\gamma)$, is one to one that is half of its values are quadratic residues, half are not. Considering the mapping $\gamma \rightarrow (x^{p-1/2} - 1,\ f(x-\gamma))$, this means that for half of the values of γ the G.C.D. will yield as a result of type b). But then if that is the case we obtain a factorisation of f(x) as follows:

if $x - \delta$ is a zero of $f(x-\gamma)$, then $x - \delta + \gamma$ is a zero of f(x).

Taking $f(x) = x^2 - a$; by a random choice of γ , one obtains a zero of f(x) in two steps, with probability one.

Instead of taking $x^{p-1/2} - 1$ as the characterization of quadratic residues, one could have taken $x^{p+1/2} + 1$ as the characterization of quadratic non residues.

Implementing the algorithm

If one implements the preceeding algorithm for signatures, one is faced with the following problems

1) Half of the messages cannot be signed, as half of the elements of F_p are quadratic non

residues.

2) The signature of a given message, is always the same allowing forgery.

3) Two different signatures for a given message allow the factorization of m.

All these weaknesses come from the deterministic nature of the above described procedure. One solution consists in randomizing the messaqe to be signed. The way to do so needs careful studying.

REFERENCE

M. Rablin: Digitalized signatures as intractable as factorisations, MIT, LCS Report 1979.

PRIMALITY TESTING – A DETERMINISTIC ALGORITHM

by S. Harari

The most ancient and simple method for testing if a number is prime or not consists in factoring n. Using the fact that a non prime has a divisor r such that $1 \leqslant r \leqslant \sqrt{n}$. We obtain a method usable for numbers up to 10^{16} on a computer. It necessitates $O(\sqrt{n})$ operations, which is large when n is big. Remarkable improvements have been made. On a pocket calculator one can factor numbers up to 10^{18} and on a large computer up to 10^{40} with $O(n^{1/4})$ algorithms.

However if one is interested in primality testing (and not to the actual factors) of a number n, there are more efficient algorithms. All these tests are based on Fermats' theorem which asserts that if n is prime and if $(a,n) = 1$ then $a^{n-1} \equiv 1 \bmod n$. We can calculate expressions of the type $a^b \pmod{c}$ in $O(\text{Log } b)$ operations, which is very fast.

Fermat's theorem yields a very fast compositeness test. Unfortunately its converse is false, therefore one cannot use it as a primality test. However, number n such that $a^{n-1} \equiv 1 \bmod n$ are quite rare. To be able to apply more severe tests than Fermat's theorem one needs a definition.

Definition: let $n > 1$ be an integer, let $n - 1 = 2$ to n with n, odd. n is said to be a strong pseudo prime of base a if either $a^{n_1} \equiv 1 \bmod n$.
or there exists $t < t_0$ such that $a^{2^t n_1} \equiv -1 \bmod n$.

Every prime number is a strong pseudo prime of base a for each a such that $(a,n) = 1$. There exists pseudo primes that are not primes. $2047 = 23.89$ is a base 2 pseudo prime.

The use of strong pseudo primes comes from the following theorem:

Theorem 1.3. Let n be a non prime. The number of integers a such that $2 \leqslant a \leqslant n-1$ and such that n is a strong pseudo prime of base a is at most $n - 1/4$.

This result yields a new type of test, called probabilistic test using the following method.

1) Choose an integer k, and chose k random integers $a_1, \ldots, a_k$.

2) If for one $i \in [1, k]$ n is not a strong pseudo prime of base a_i, n is composite. Otherwise we shall declare n to be prime with probability of error less than 4^{-k}.

Adleman's test

This test declares that a number is prime or composite, when the number is prime or composite, without using improved theorems like the Extend Riemann Hypothesis.

The most simple version of this test is a probabilistic version, in another sense than the preceeding one: if n is prime, the algorithm will prove it to be prime without any error. The only probabilistic aspect is that it may happen that the algorithm will fail in determining whether n is prime or not. However if n is composite it is practically certain that the Miller Rabin test will show it, therefore Adleman's test is to be used after using Miller Rabin's test.

Execution time of this algorithm is $0((\mathrm{Log}\, n)^{c \log\log\log n})$, where C is an effective constant.

Let us introduce some notations.

Let n be a fixed prime to be tested. Let p and q be two numbers, prime, such that p^k $(q-1)$ and $(pq, n) = 1$. Let ζ_{pk} (resp. ζ_q) be a p th root of unity (resp. q-th root of unity). Let $< \zeta_{p^k} >$ be the cyclic group of all p^k-th root of unity.

$G = \mathrm{Gal}(Q(\zeta_{p^k})(Q) = \{\sigma_x \mid 1 \leqslant x \leqslant p^k - 1)\ p \chi x\}$ where σ_x is the automorphism of $Q(\zeta_{p^k})$ which send ζ_{p^k} to $\zeta^x_{p^k}$. Let $\mathscr{P} = \{\gamma \in Z[G]/\ \zeta^{\gamma}_p = 1\}$. $\mathscr{P}$ is a prime ideal generated by all $\sigma_x - x$ for $x \in Z - q\,Z$ and by p.

Let χ be the character of order p^k and conductor q.

If $p \geqslant 3$ then $\chi(-1) = 1 \quad (q-1)/2^k$

If $p = 2$ then $\chi(-1) = (-1)$.

The Gauss sum defined by:

$$\tau(\chi) = \sum_{x \in F_q^*} \chi(x)\ \zeta^x_q \quad \in \quad Z[\zeta_{pk}, \zeta_q]$$

satisfies the condition $\tau(\chi)\ \ \tau(\chi) = \chi(-1)q \cdot \tau(\chi)$ is invertible mod n since $(q,n) = 1$.

Let β be an element of $Z\ [G]$ such that $\beta \in \mathscr{P}$.

The following condition $(*\ \beta)$

$(*\ \beta)$ There exists $\eta(\chi) \in < \zeta_{pk} >$ such that

$$\tau(\chi)^{\beta(n-\sigma_n)} \equiv \eta(\chi)^{-\beta n} \mod n$$

has the following property.

Lemma: If n is prime $(*\ \beta)$ is true with $\eta(\chi) = \chi(n)$.

The basic idea behind the test is that if (* β) is true for "enough" characters χ (with various p and q) then one has enough elements to decide if n is prime. This leads to the fundamental theorem.

Theorem: Let t be an even integer such that

$$e(t) = 2 \prod_{\substack{q \text{ prime} \\ (q-1)|t}} \cdot q^{v_q(t)+1} \quad \text{be such that } (n, t\ e(t)) = 1.$$

For all pairs (p, q) of prime numbers with $(q-1)|t$ and $P^k \| q-1$ choose a character χ_{pq} of order p^k and conductor q (one can take $\chi(g_q^a) = \zeta_{pk}^q$ if g_q is a primitive root mod q). Suppose that 1) for each pair (p, q) as above χ_{pq} verifies (* β) for a certain B ϵ P.

1) for each pair (p, q) as above χ_{pq} verifies (* β) for a certain $B \notin \mathscr{P}$
2) for all $p|t$, either $p \geqslant 3$ and $n^{p-1} \equiv 1 \bmod p^2$ or there exists q such that
a) $\eta(\chi_{pq})$ is a primitive p^k th root of unity, where $k = v_p(k-1)$
and
b) $p \geqslant 3$
or $p = 2$ $k = 1$ and $n \equiv 1 \bmod 4$
or $p = 2$ $k \geqslant 2$ and $q^{(n-1)/2} \equiv -1 \bmod n$.

Then for all divisors r of n there exists $i \in [0, t-1]$ such that $r \equiv n^i \bmod (e(t))$.
This theorem leads to the following test.

Deterministic primality test (Adleman, Rumely, Pomerance, Lenstra)

1) This part depends only on the size of n, and not of its value, and can be satisfied for all n of comparable size).
a) choose t such that $e(t) > \sqrt{n}$. For $n < 4.10^{313}$ one can use table [1]. The choice of the smallest possible t is not always the best one.
b) for all $q|e(t)$ choose g_q a primitive root modulo q. For all $p|q-1$ let $k = v_p(k-1)$ and χ_{pq} the character defined by $\chi_{pq}(g_q^x) = \zeta_{pk}^x$.
2) Check that $(n, t\ e(t)) = 1$.
3) For all pairs p, q. Check that χ_{pq} satisfies (* β) for a $\beta \notin \mathscr{P}$ (for instance $\beta = 1$).
If it is not the case n is composite
4) Check that for all prime $p|t$ one of these conditions is satisified.
a) $p \geqslant 3$ and $n^{p-1} \equiv 1 \bmod p^2$.
b) $p \geqslant 3$ and $\eta(\chi_{pq})$ is a **primitive** p^k th root of 1; for at least one $q|e(t)$ where $k =$
$= v_p(q-1)$.
c) $p = 2$, $n \equiv 1 \bmod 4$ and $\eta(\chi_{2q}) = -1$ for at least a $q|e(t)$ such that $v_2(q-1) = 1$

(or $q \equiv 3 \mod 4$).

d) $p = 2$ and there exists a $q | e(t)$, $q \equiv 1 \mod 4$ such that $\eta(\chi_{2q})$ be a primitive 2^k th root of 1 ($k = v_2(q-1)$) and furthermore $q^{n-1/2} \equiv -1 \mod n$.

If these conditions are true for all $p | t$, go to 6. If not.

5) For all $p | t$ where 4) was not verified: choose a certain number of $q \chi e(t)$ such that $q \equiv 1 \mod p$ and $(q,n) = 1$; choose g a primitive root mod q, let χ_{pq} be the character defined by $\chi_{pq}(g_q^X) = \zeta_{pk}^x$ with $k = v_p(q-1)$. Check that χ_{pq} satisfies ($* \beta$). If not n is composite. Check that one condition of 4 is satisfied for at least one q. If it is not the case the test fails.

6) For $0 < i < t$ compute $R(n^i, e(t))$, remainder of n^i divided by $e(t)$. If there exists an i such that $R(n^i, e(t)) \neq 1$, n and divides n, n is composite otherwise n is prime.

Remark: step 5 has been added to make the probability of failure as small as possible. When n is big one can prove that for each p_i there exists $q \equiv 1 \mod p$ such that the conditions in 4) are satisfied.

The execution time of this last test depends on the size of t. The time given as $O((\text{Log}\, n)^{c \text{ Log Log Log}\, n})$ results from the following theorem.

Theorem: Odlyzko-Pomerance. There exists t without square factors verifying $t < (\text{Log}\, n)^{c \text{ Log Log Log}\, n}$ (where c is an effective constant) and such that $e(t) > \sqrt{n}$.

This bound has a theoretical interest. In practice for number $n < 4.10^{313}$ table 1 is enough. The largest necessary t is then 166320 which is not very large.

REFERENCE

[1] Test de primalité d'après, Adleman, Rumely, Pomerance, Lenstra. H. Cohen. Université scientifique et medicale de Grenoble.

[2] Lenstra, Tests de primalité, Seminaire Bourbaki, Juin 1981.

[3] Rabin, Probabilistic algorithms for testing primality. J. Number theory 12 (1980) pp. 128-138.

[4] Williams, Primality testing on a computer, Ars combina toria, 5 (1978).

TABLE 1

For each integer t, let $e(t) = 2 \prod_{\substack{q \text{ prime} \\ (q-1)|t}} \cdot q^{v_q(t)+1}$.

In the first column are indicated the t's for which $\forall u < t,\ e(u) < e(t)$, as well as their factorisation. In the second column are indicated lower bounded approximations to $e(t)^2$.

t	$(e(t))^2$	t	$(e(t))^2$
$2 = 2$	5,7600000 E 002	$15\,120 = 2^4.3^3.5.7$	5,0831601 E 158
$4 = 2^2$	5,7600000 E 004	$25\,200 = 2^4.3^2.5^2.7$	1,7575997 E 179
$6 = 2.3$	2,5401600 E 005	$30\,240 = 2^5.3^3.5.7$	3,6417363 E 191
$12 = 2^2.3$	4,2928704 E 009	$42\,840 = 2^3.3^2.5.7.17$	2,9544609 E 202
$24 = 2^3.3$	1,7171481 E 010	$45\,360 = 2^4.3^4.5.7$	1,5723798 E 206
$30 = 2.3.5$	2,9537234 E 010	$55\,440 = 2^4.3^2.5.7.11$	2,4216182 E 213
$36 = 2^2.3^2$	1,9094176 E 016	$60\,480 = 2^6.3^3.5.7$	1,6174651 E 232
$60 = 2^2.3.5$	4,6436150 E 019	$75\,600 = 2^4.3^3.5^2.7$	3,8056040 E 233
$72 = 2^3.3^2$	4,0701147 E 020	$85\,680 = 2^4.3^2.5.7.17$	6,2539126 E 258
$108 = 2^2.3^3$	2,0417212 E 021	$100\,800 = 2^6.3^2.5^2.7$	7,8670318 E 268
$120 = 2^3.3.5$	3,1223667 E 023	$110\,880 = 2^5.3^2.5.7.11$	4,4490778 E 274
$144 = 2^4.3^2$	4,7050525 E 023	$128\,520 = 2^3.3^3.5.7.17$	7,2908267 E 290
$180 = 2^2.3^2.5$	6,7665379 E 030	$131\,040 = 2^5.3^2.5.7.13$	6,2430632 E 303
$240 = 2^4.3.5$	2,0964081 E 031	$166\,320 = 2^4.3^3.5.7.11$	4,8871405 E 313
$360 = 2^3.3^2.5$	2,4245991 E 038		
$420 = 2^2.3.5.7$	1,4074436 E 041		
$540 = 2^2.3^3.5$	1,5552318 E 046		
$720 = 2^4.3^2.5$	1,6279155 E 046		
$840 = 2^3.3.5.7$	7,4725932 E 049		
$1\,008 = 2^4.3^2.7$	8,5368948 E 050		
$1\,080 = 2^3.3^3.5$	5,5727372 E 053		
$1\,200 = 2^4.3.5^2$	1,0212596 E 058		
$1\,260 = 2^2.3^2.5.7$	1,3170657 E 062		
$1\,620 = 2^2.3^4.5$	6,4271819 E 063		
$1\,680 = 2^4.3.5.7$	7,2757866 E 066		
$2\,016 = 2^5.3^2.7$	5,9203218 E 067		
$2\,160 = 2^4.3^3.5$	3,2760144 E 073		
$2\,520 = 2^3.3^2.5.7$	2,3683136 E 081		
$3\,360 = 2^5.3.5.7$	1,4010401 E 084		
$3\,780 = 2^2.3^3.5.7$	2,4917603 E 088		
$5\,040 = 2^4.3^2.5.7$	2,3476327 E 104		
$6\,480 = 2^4.3^4.5$	9,5661076 E 104		
$7\,560 = 2^3.3^3.5.7$	2,5615126 E 115		
$8\,400 = 2^4.3.5^2.7$	2,6638720 E 119		
$10\,080 = 2^5.3^2.5.7$	1,8391306 E 128		
$12\,600 = 2^3.3^2.5^2.7$	9,7437932 E 137		

Communication in the Presence of Jamming-
An Information-Theoretic Approach

Robert J. McEliece
California Institute of Technology
Pasadena, California 91125 U.S.A.

1. Introduction. In traditional information theoretic studies, the channel is entirely passive, though possibly quite complex probabilistically. However, it sometimes happens in practice that the channel is partially controlled by an adversary (the Jammer) whose goal is to do everything in his power to make communication difficult. This possibility leads to a whole host of interesting mathematical and engineering problems, and in this paper we will study a few of these. In the next section, we will introduce a two-person zero-sum game with mutual information as the payoff function. The first player (the Communicator) wants to maximize this function, and the second player (the Jammer) wants to minimize it. Since mutual information is convex-concave in just the right way, a generalization of Von Neumann's minimax theorem

turns out to guarantee the existence of jointly optimal saddlepoint strategies for the players. We show that these strategies are memoryless for both players.

In Section 3, motivated by the theorem of memoryless saddlepoint strategies, we briefly discuss the design of an algebraic pseudo-random data scrambler. This subject has only recently been studied to any depth in the open literature, and we hope our simple example will stimulate further research in this interesting subject.

In Section 4 we will give several explicit examples of channels with jamming, and describe the corresponding saddlepoint strategies.

In Section 5 we will present what is by now a classic result in this area: Houston's derivation of the optimal partial-band jammer vs. noncoherent FSK modulation. We will show that an uncoded communicator is at a hopeless disadvantage against such a jammer, but that with coding the jammer can be essentially neutralized.

2. A Game-Theoretic Formulation. Our channel model has input alphabet A and output alphabet B. There are two players. Player 1, also called the Communicator, chooses the input X to the channel. X is just an A-valued random variable. A is required to lie in a certain set S, called the set of allowable inputs. Player II, also called the Jammer, chooses for each $a \in A$ a B-valued random variable Y^a, which represents the channels' output, given that a is the input. This collection $\{Y^a\}$, $a \in A$, which for brevity we denote by Z, is required to lie in a certain set T, its set of allowable channels. X and Z are called the

strategies adopted by the players. The situation can be depicted as in Fig. 1. In Fig. 1 Y denotes the channel output if X is the Communicator's strategy and Z is the Jammer's strategy.

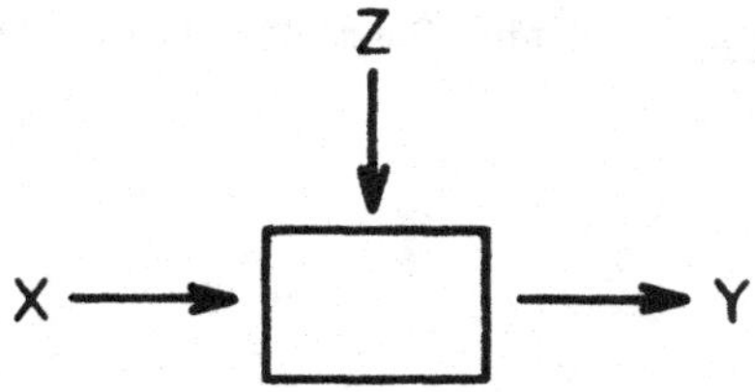

Figure 1. An abstract model for jamming.

The payoff function for this game, denoted by $\phi(X,Z)$, is defined to be the mutual information $I(X;Y)$ between the random variables X and Y:

$$\phi(X,Z) = I(X;Y).$$

Of course the Communicator's goal is to make ϕ as large as possible and the Jammer's goal is to make ϕ as small as possible. Thus associated with this game are two programs:

Program I (Communicator's Program):

$$C' = \sup_{X \in S} \inf_{Z \in T} \phi(X,Z) \tag{2.1}$$

Program II (Jammer's Program):

$$C'' = \inf_{Z \in T} \sup_{X \in S} \phi(X,Z) . \tag{2.2}$$

A strategy $X_0 \in S$ such that

$$\inf_{Z \in T} \phi(X_0, Z) = C' \tag{2.3}$$

is called an <u>optimal strategy</u> for the Communicator. The significance is that (2.3) implies

$$\phi(X_0, Z) \geq C' \tag{2.4}$$

for all allowable channels Z. Hence, if the communicator chooses the input X_0, he is guaranteed a payoff of at least C', regardless of the Jammer's strategy. Sometimes no such optimal strategy exists, and the Communicator must be satisfied with an <u>ε-optimal strategy</u>, i.e. an $X_0 \in S$ such that for a given choice of $\epsilon > 0$

$$\phi(X_0, Z) \geq C' - \varepsilon \tag{2.5}$$

for all allowable noises Z.

Similarly, we define optimal strategies and ε-optimal strategies for the Jammer, and in place of (2.3), (2.4), and (2.5), we have for a given $Z_0 \in T$

$$\sup_{X \in S} \phi(X, Z_0) = C'', \tag{2.6}$$

$$\phi(X, Z_0) \leq C'' \qquad \text{for all } X \in S, \text{ and} \tag{2.7}$$

$$\phi(X, Z_0) \leq C'' + \varepsilon \qquad \text{for all } X \in S \tag{2.8}$$

(depending on whether the strategy Z_0 is optimal or ε-optimal).

If it happens that $C' = C''$, then combining (2.4) and (2.7), we have

$$\phi(X_0,Z_0) = C' = C'', \text{ and} \tag{2.9}$$

$$\phi(X,Z_0) \le \phi(X_0,Z_0) \le \phi(X_0,Z) \tag{2.10}$$

for every choice of allowable X and Z. If (2.9) holds, which is equivalent to (2.10), the common value, denoted by C, is called the value of the game. The pair (X_0,Z_0) of optimal strategies is called a saddlepoint. In absence of other information, the Coder will want to play strategy X_0, and the Jammer will want to play Z_0.

It follows immediately from the definitions that $C' \le C''$, and it is easy to give examples such that $C' < C''$. However, since ϕ is convex $\cap$ in X and convex $\cup$ in Z ([8], Theorems 1.6 and 1.7), if S and T are compact and convex in an appropriate sense, it will follow from a generalization of Von Neumann's minimax theorem ([1]) that a saddlepoint exists. Rather than state a precise abstract existence theorems, we will let the subject of saddlepoints rest until Section 4, where we will explicitly compute several. In the remainder of this section we will discuss certain multi-dimensional generalizations.

We now generalize the game by allowing the players to adopt n-dimensional strategies. We first need to appropriately generalize the notion of admissibility to n-dimensional strategies.

Definition: If $X_1,X_2,\ldots,X_n$ are random variables, with cumulative distributions $F_1,F_2,\ldots,F_n$, we define their uniform mixture to be the random variable X with c.d.f.

$$\hat{F} = \frac{1}{n} \sum_{k=1}^{n} F_k .$$

(Informally, $\hat{X}$ is equal to X_k with probability $\frac{1}{n}$.)

We now say that an n-dimensional random input $\underline{X} = (X_1, X_2, \ldots, X_n)$ is an admissible strategy for the Communicator, and write $\underline{X} \in S$, if the uniform mixture of the components X_i lies in S. We might call this an average admissibility criterion.

The definition of average admissibility for the Jammer is a bit more complicated. An n-dimensional channel is a collection $\underline{Z} = \{\underline{Y}^{\underline{a}}\}$, of B^n-valued random variables, one for each $\underline{a} \in A^n$. If the components of $\underline{Y}^{\underline{a}}$ are denoted by $(Y_1^{\underline{a}}, Y_2^{\underline{a}}, \ldots, Y_n^{\underline{a}})$, and if for each $\underline{a} \in A^n$ we denote by $Y^{\underline{a}}$ the uniform mixture of $Y_1^{\underline{a}}, Y_2^{\underline{a}}, \ldots, Y_n^{\underline{a}}$, we say that $\{\underline{Y}^{\underline{a}}\}$ is admissible, and write $\underline{Z} \in T$, if $\{Y^{\underline{a}}\} \in T$.

Given the definitions, we denote the values of the n-dimensional generalized programs by C_n' and C_n'' :

Program I_n:

$$C' = \sup_{\underline{X} \in S} \inf_{\underline{Z} \in T} \frac{1}{n} \phi(\underline{X}, \underline{Z}).$$

Program II_n:

$$C'' = \inf_{\underline{Z} \in T} \sup_{\underline{X} \in S} \frac{1}{n} \phi(\underline{X}, \underline{Z}).$$

It is surprising that allowing this extra generality does not alter the values of the programs:

Theorem 2.1 $C_n' = C'$ and $C_n'' = C''$.

Proof: We shall prove only $C_n' = C'$, since the other result is proved in almost exactly the same way. Let $X \in S$ be any one-dimensional admissible strategy for the Communicator, and let $\underline{Z} \in T$ be any n-dimensional admissible strategy for the Jammer. Then if $\underline{X} = (X_1, X_2, \ldots, X_n)$ is a random vector consisting of n independent copies of X, we have:

$$\frac{1}{n} \phi(\underline{X},\underline{Z}) \geq \frac{1}{n} \sum_{i=1}^{n} \phi(X,Z_i) \qquad \text{([8], Theorem 1.8)}$$

$$\geq \phi(X;\hat{Z}) \qquad \text{([8], Theorem 1.7)},$$

where $\hat{Z}$ is the uniform mixture of the components of $\underline{Z}$. Hence, since $Z \in T$,

$$\inf_{\underline{Z} \in T} \frac{1}{n} \phi(\underline{X},\underline{Z}) \geq \inf_{Z \in T} \phi(X,Z),$$

$$\sup_{\underline{X} \in S} \inf_{\underline{Z} \in T} \frac{1}{n} \phi(\underline{X},\underline{Z}) \geq \sup_{X \in S} \inf_{Z \in T} \phi(X,Z),$$

and so $C_n' \geq C'$. On the other hand, let $\underline{X} \in T$ be arbitrary. Then if $Z = (Z_1, Z_2, \ldots, Z_n)$, where the Z_i's are independent copies of Z (so $\underline{Z}$ is the n-th memoryless extension of Z), we have

$$\frac{1}{n} \phi(\underline{X},\underline{Z}) \leq \frac{1}{n} \sum_{i=1}^{n} \phi(X_i,Z) \qquad \text{([8], Theorem 1.9)}$$

$$\leq \phi(\hat{X},Z) \qquad ([8], \text{Theorem } 1.6),$$

where $\hat{X}$ is the uniform mixture of the components of $\hat{X}$. Hence, since $X \in S$, we have

$$\inf_{Z \in T} \frac{1}{n} \phi(\underline{X},\underline{Z}) \leq \inf_{Z \in T} \phi(\hat{X},Z)$$

$$\sup_{X \in S} \inf_{Z \in T} \frac{1}{n} \phi(\underline{X},\underline{Z}) \leq \sup_{X \in S} \inf_{Z \in T} \phi(X,Z),$$

and so $C_n' \leq C'$. Thus $C_n' = C'$, as asserted. QED.

Discussion: The importance of Theorem 1 is that it shows that optimal strategies are memoryless (or, more precisely, that there exist memoryless optimal strategies.) The interpretation of this for the Jammer is simpler, and so we discuss it first.

If the Jammer presents to the Communicator a memoryless channel whose statistics are those of an optimal strategy for Program II, then the Communicator cannot possibly communicate reliably at rates higher than C'', no matter what strategy he adopts. (The proof of this follows from Theorem 5.1(a) of [8], with the modification that $I(\underline{X};\underline{Y}) \leq \sup_{X \in S} I(\underline{X},\underline{Y}) = C_n'' = C''$ when the Jammer uses his optimal strategy.) Thus C'' is the smallest channel capacity the Jammer can present to the communicator, under any circumstances, and a memoryless strategy affords this channel. No admissible Jamming strategy, memoryless or not, can do better.

We now take the Communicator's viewpoint. If the Jammer's channel is memoryless (which according to our previous discussion it ought to be), then according to a theorem of Blackwell, Breiman, and Thomasian, ([19] Theorem 4.3.1), for any $R < C'$, it is possible to construct a code of rate R together with a decoding rule, such that the decoder error probability is negligible, regardless of the channel statistics $Z \epsilon T$. The proof is by random coding, the length-n codeword coordinates chosen according to an optimal strategy for the Communicator (Program I_n), which, by Theorem 2.1, may be taken as i.i.d. random variables. We do not know of a similar result for an arbitrary channel $\underline{Z} \epsilon T$, but as a practical matter if a scrambling protocol is adopted by the encoder and decoder, any realistic channel will be rendered essentially memoryless, and the BBT theorem applies. Furthermore ([19], Theorem 4.4.1), the Coder cannot construct a code of rate above C' which gives small error probability even for all memoryless channels $\underline{Z} \epsilon T$, and so C' is the channel capacity, from the Coder's viewpoint.

In the next section we will briefly discuss the design of the scramblers needed by the Communicator to render a complex Jammer memoryless.

3. A Pseudo-Random Scrambler. In the last section we noted the desirability, under some circumstances, of data scrambling vs. a non-memoryless jammer. In this section we will describe one class of scramblers, most effectively implementable with one RAM (random access memory.)

The essence of our scrambler is shown in Fig. 2.

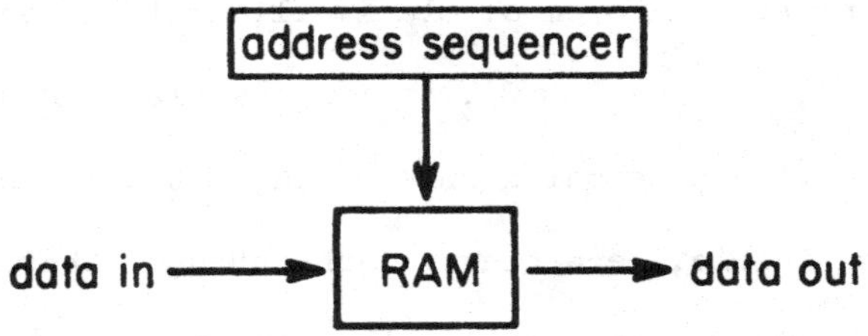

Figure 2. A One-RAM Scrambler.

On each clock cycle, the address sequencer provides a new address, one bit (or larger character) is read out from this address, and a new input character is read into that same address. In this way a variable delay is introduced into the data stream, the details of which depend on the details of the address sequencer. This idea is discussed at length in [2], but here let us describe one such implementation, based on an address sequencer which is a maximal-length linear shift register sequence.

To illustrate, let us consider an 8 x 1 RAM with memory addresses labeled 000,001,...,111. The address sequencer is a 4-stage MLSR described by the primitive feedback polynominal D^4+D+1. (Figure 3).

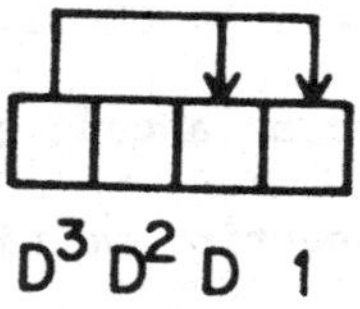

Figure 3. MLSR Address Sequencer

If the initial contents of the shift register in Figure 3 are 0001, then because of the primitive logic, the contents of the SR will cycle through all 15 4-bit nonzero states in the order 0001, 0010, 0100, 1000, 0011, 0110, 1100, 1011, 0101, 1010, 0111, 1110, 1111, 1101, 1001, and then repeat periodically. The idea is now to use the leftmost three bits of the shift register to generate addresses in the 8 x 1 RAM. Thus the addresses generated this way (in decimal) will be 0, 1, 2, 4, 1, 3, 6, 5, 2, 5, 3, 7, 7, 6, 4, and then repeated periodically.

Let us now suppose that the data to be scrambled is the sequence A, B, C, D, E, F, ... The character 'A' will be written into memory location 0, and not read out until 0 appears in the address sequence again, viz. 15 clock cycles later. 'B' will be written into location 1 and read out again 3 clock cycles later; 'C' goes into location 2, and is read out 6 cycles later, etc. In this way a variable delay is introduced into the data stream, as shown below:

MLSR Contents	RAM Address	Delay Introduced
0001	0	15
0010	1	3
0100	2	6
1000	4	11
0011	1	12
0110	3	5
1100	6	7
1011	5	2
0101	2	9
1010	5	13
0111	3	10
1110	7	1
1111	7	14
1101	6	8
1001	4	4

We notice a remarkable phenomenon: each possible delay in the range [1,15] occurs exactly once! If the delay is viewed as a random variable, this random variable is uniformly distributed on [1,15]. This is of course no accident. The same thing happens for any $2^m \times 1$ RAM controlled by an (m+1)-bit MLSR address sequencer. This can be proved as follows.

Let us view the MLSR contents as the vector representation of the powers of a primitive sort α in the finite field GF (2^{m+1}). From this viewpoint the MLSR cycles through $1,\alpha,\alpha^2,\ldots,\alpha^{2^{m+1}-2}$, and then repeats. The address corresponding to a given element α^j is just the first (high-order) m bits in the vector representation of α^j. The element stored in this address will undergo a delay of d cycles if and only if α^{j+d} gives the same address as α^j, i.e. if $\alpha^{j+d} = \alpha^j$ or $\alpha^j + 1$. Now $\alpha^{j+d} = \alpha^j$ requires $\alpha^d = 1$, which forces d to be a multiple of $2^{m+1} - 1$. On the other hand for any $1 \le d \le 2^{m+1} - 1$, the equation $\alpha^{j+d} = \alpha^j + 1$ has a unique solution for α^j, viz. $\alpha^j = (\alpha^d + 1)^{-1}$. Thus for any d in this range, there is a unique position in the address sequence (mod 2^{m+1} -1) which will cause a delay of d.

It follows from this theorem, e.g. that if we used a 64K RAM (m = 2^{16}), by controlling the addresses with a 17-stage MLSR (e.g. $D^7 + D^3 + 1$), we would obtain a scrambler which introduced a pseudo-random delay uniformly distributed in (1,2,3,...,131071). We pose as a research problem the analyses of the delays introduced if nonlinear SL logic is used to control the address sequencer.

4. Some Saddlepoints. In this section we will investigate in detail three special cases of the general problem outlined in Section 2. All of these cases deal with additive noise (Figure 4).

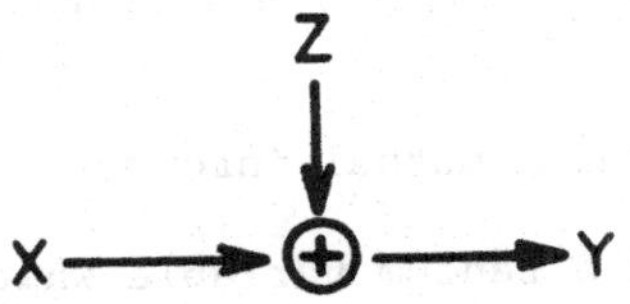

Figure 4. Additive Noise Model.

In Fig. 4, X is the Communicator's signal, and Z is the Jammer's noise. The receiver observes the sum of X and Z, denoted by Y, or perhaps a function of Y. In view of Theorem 2.1, above, we consider only the memoryless version of this problem, viz. X and Z are independent one-dimensional random variables. The restrictions on X and Z are normalized mathematical abstractions of average power constraints:

$$E(X^2) \leq A \tag{4.1}$$

$$E(Z^2) \leq 1 \tag{4.2}$$

The three problems to be described in this section can now be specified by Figure 4, the constraints (4.1) and (4.2), and the following three objective functions:*

Objective Function 1: $\phi(X,Z) = I(X;Y)$ (4.3)

Objective Function 2: $\phi(X,Z) = I(\mathrm{sgn}(X);\mathrm{sgn}(Y))$ (4.4)

Objective Function 3: $\phi(X,Y) = I(X;\mathrm{sgn}(Y))$ (4.5)

Here $I(X;Y)$ denotes the Shannon mutual information between X and Y. Here "sgn(X)", the sign of X, is a random variable which is +1 if X is positive, -1 if X is negative, and equally likely to be ±1 when X is zero. If the receiver uses a hard limiter, it will be sgn(Y) and not Y that is observed. Hard limiters are often encountered in practice. The importance of sgn(X) is less apparent. However, consider a communication system in which the data to be transmitted are binary, e.g. ±1. Suppose that in order to thwart the Jammer, the Communicator transmitted this binary data using random amplitudes. By this we mean a strategy in which the transmitted signal X has the same sign as the data bit, but whose absolute value is a random variable, chosen by the Communicator. Then the transmitted information is the random variable sgn(X), and the appropriate payoff function is the mutual information between sgn(X) and the received random variable.

In each of the three cases, we denote the value of the game by $C^*(A)$. Here are the results (proofs follow):

*There is an obvious Objective Function 4, viz $\phi(X,Z) = I(\mathrm{sgn}(X);Y)$, but we have not at present solved this case.

Theorem 4.1 (Shannon, Blachman, Dobrushyn) For objective function 1,

$$C^*(A) = \frac{1}{2} \log (1 + A). \tag{4.6}$$

Both the Communicator and the Jammer have unique optimal strategies given by

$$X_0 = N(0,A) \tag{4.7}$$

$$Z_0 = N(0,1), \tag{4.8}$$

where $N(\mu,\sigma^2)$ denotes a normal random variable with mean μ and variance σ^2.

Theorem 4.2. For objective function 2,

$$C^*(A) = \log 2-H(\epsilon^*), \tag{4.9}$$

where $H(x) = -x\log x-(1-x)\log(1-x)$ is the entropy function, and $\epsilon^* = \epsilon^*(A)$ is given by

$$\epsilon^*(A) = \begin{cases} \frac{1}{4A}, & A \geq 1 \\ \frac{1}{2} - \frac{A}{4}, & A \leq 1. \end{cases} \tag{4.10}$$

The optimal strategies X_0 and Z_0 in this case are best described in terms of the random variables X_0^2, Z_0^2, it being understood that the signs of X_0 and Z_0 are random, and equally likely to be plus or minus.

For $A \geq 1$: X_0^2: uniform on $[0,2A]$ (4.11)

Z_0^2: 0 with probability $1-1/A$

uniform on $[0,2A]$ with probability $1/A$

For $A \leq 1$: X_0^2: 0 with probability $1-A$ (4.12)

uniform on $[0,2]$ with probability A

Z_0^2: uniform on $[0,2]$.

To describe the optimal strategies for objective function 3, we introduce the following class of distribution functions $\{F_\mu : \mu \in (0, \log 2)\}$. Choose any $\mu \in (0, \log 2]$. Observe that since H is strictly increasing on $(0, \frac{1}{2}]$, there exists a unique $x \in (0, \frac{1}{2}]$ such that $H(x) = \mu$.

For any $\mu \in (0, \log 2]$ define

$$\lambda = \lambda(\mu) = 2\int_0^\mu H^{-1}(y)dy \tag{4.13}$$

$$= 2\,\mu x - 2\int_0^x H(t)dt$$

$$= 2\,\mu x - x - (1-x)^2 \log (1-x) + x^2 \log x,$$

(natural logarithms.)

where $x = H^{-1}(\mu)$. For any such choice of μ, let F be the unique distribution function symmetric about zero satisfying

$$H(F_\mu(z)) = \begin{cases} -\lambda z^2 + \mu & \text{if } |z| \leq \sqrt{\mu/\lambda} \\ 0 & , \text{if } |z| > \sqrt{\mu/\lambda}. \end{cases} \tag{4.14}$$

When $\mu = \log 2$, F_μ is differentiable everywhere, but for $\mu < \log 2$, F_μ is differentiable everywhere except at zero where it has a jump discontinuity of height $1 - 2H^{-1}(\mu)$. The derivative of F_μ for $\mu = \log 2$ is shown in Fig. 5.

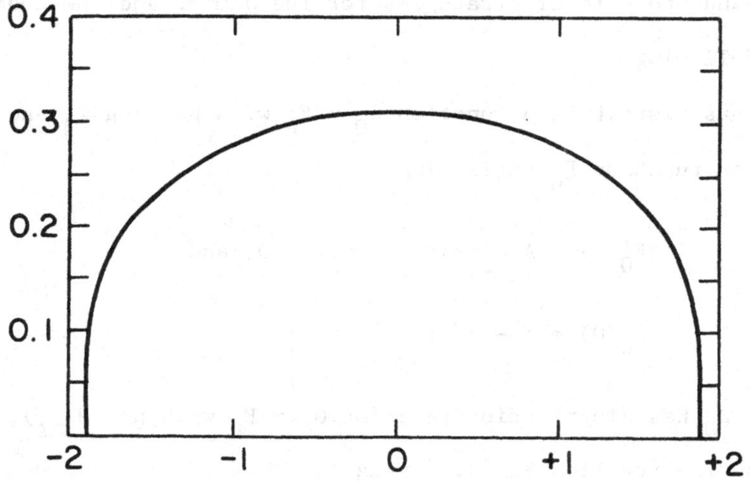

Figure 5. The function $F_{\log 2}(x)$. (The x-intercepts are $\pm \sqrt{\log 2/(\log 2-1/2)}$

A simple calculation shows that for each $\mu \in (0, \log 2)$

$$\int z^2 dF_\mu(z) = 1.$$

Theorem 4.3. For objective function 3,

$$C^*(A) = \begin{cases} A(\log 2 - \frac{1}{2}) & \text{for} \quad A \le 1, \text{ and} \\ (\log 2 - \frac{1}{2}) + \frac{x}{2}\log x - \frac{(1-x)^2}{2x}\log(1-x) & \text{if } A \ge 1. \end{cases} \tag{4.15}$$

where $x = \frac{1}{2A}$; and the optimal strategies for the Jammer and the Coder are given by the following:

For $A < 1$, Z_0 has distribution function $G_0 = F_\mu$ with $\mu = \log 2$, and X_0 has distribution function F_0 satisfying

$$F_0'(z) = A\, G_0'(z) \qquad \text{for } z \ne 0, \text{ and}$$

$$F_0(0) = 1 - A/2;$$

and for $A \ge 1$, Z_0 has distribution function $G_0 = F_\mu$ with $\mu = H(\frac{1}{2A})$, and X_0 has distribution function F_0 satisfying

$$F_0'(z) = A_0\, G_0'(z) \qquad \text{for } z \ne 0, \text{ and}$$

$$F_0 \text{ is continuous at } z \ne 0.$$

(It is easy to show that $E\, X_0^2 = A$.)

In Figure 6 we have plotted the functions $C^*(A)$, measured in bits, for the three objective functions, for the range $0 \le A \le 4$. Note that for $A = 1$, both players have the same optimal strategy, the one whose density function is shown in Figure 5.

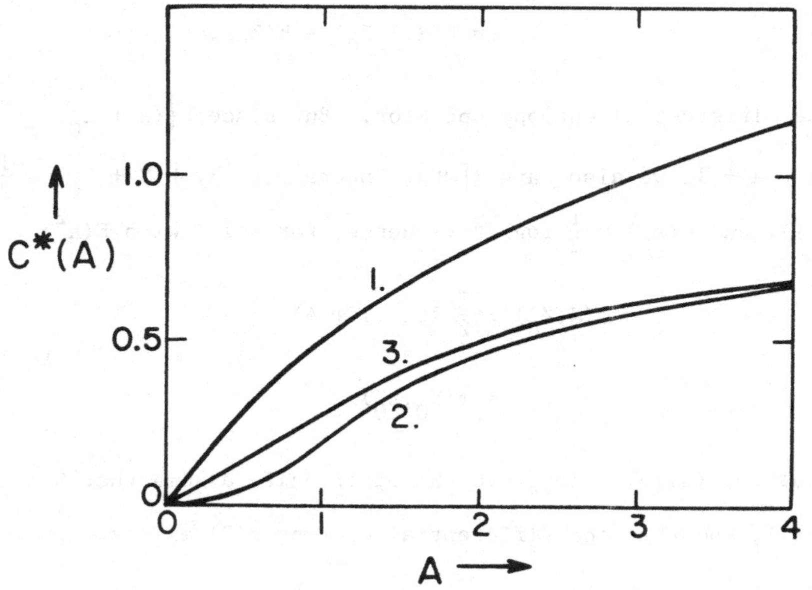

Figure 6. The function C*(A) for objective functions 1, 2, and 3.

The rest of this section is devoted to the proofs of Theorems 4.1, 4.2, and 4.3. The details differ, but in each of the proofs the technique employed is to establish the pair of inequalities

$$\phi(X,Z_0) \leq \phi(X_0,Z_0) \leq \phi(X_0,Z), \tag{4.16}$$

where X_0 and Z_0 are the saddlepoint strategies given in the Theorem statements, and X and Z are arbitrary random variables satisfying $E(X^2) \leq A$, $E(Z^2) \leq 1$.

Proof of Theorem 4.1. Let Z_0 be a N(0,1) random variable. Then for any X which is independent of Z_0, we have ([8], Theorem 1.10)

$$\phi(X,Z_0) = I(X;X + Z_0)$$

$$= h(X + Z_0) - h(Z_0),$$

where h is the differential entropy operator. But since $E\left((X + Z_0)^2\right) = E(X^2) + E(Z^2) \leq A + 1$, we also have ([8], Theorem 1.11), $h(X + Z_0) \leq \frac{1}{2} \cdot \log 2\pi e\ (A + 1)$, and $h(Z_0) = \frac{1}{2} \log 2\pi e$. Hence, for all X with $E(X^2) \leq A$,

$$\phi(X,Z_0) \leq \frac{1}{2} \log (1 + A)$$

$$= \phi(X_0,Z_0). \tag{4.17}$$

This proves half of (4.16). To prove the other half, assume that Z satisfies $E(Z^2) \leq 1$, and that the differential entropy h(Z) exists. Then we have as before,

$$\phi(X_0,Z) = h(X_0 + Z) = h(Z).$$

But $h(X_0 + Z) \geq \frac{1}{2} \log (2^{2h(X_0)} + 2^{2h(Z)})$, by a difficult theorem of Shannon ([15, Appendix 6], [4]). Hence

$$\phi(X_0,Z) \geq \frac{1}{2} \log (2^{2h(X_0)-2h(Z)} + 1)$$

$$\geq \frac{1}{2} \log (1 + A),$$

since $h(X_0) = \frac{1}{2} \log 2\pi eA$, $h(Z) \leq \frac{1}{2} \log 2\pi e$. This proves that

$$\phi(X_0,Z) \geq \frac{1}{2} \log (1 + A)$$

$$= \phi(X_0,Z_0),$$

provided $h(Z)$ exists. If Z is not smooth enough for $h(Z)$ to exist, a similar proof can be given by considering the sequence $\{Z_n\}$, where $Z_n = Z + W_n$, where W_n is a $N(0,\frac{1}{2})$ independent from X_0 and Z. (The details of this argument can be found in [5].)

Proof of Theorem 4.2. We will give proof only in the case $A \geq 1$. The proof for $A < 1$ is similar and is omitted.

We begin by proving that $\phi(X,Z_0) \leq \phi(X_0,Z_0)$, where X_0 and Z_0 are as described in (4.11). For a fixed Z_0, $\phi(X,Z_0)$ is a convex ∩ function of X, ([8], Theorem 1.6) and so if $\hat{X}$ is the "symmetrized" version of X, $E(\hat{X}^2) = E(X^2)$ and $\phi(\hat{X}, Z_0) \geq \phi(X,Z_0)$. We therefore assume that X is already symmetric.

If X is symmetric, then since Z_0 is also, the channel connecting sgn(X) and sgn(Y) is a binary symmetric channel. Its crossover probability $\epsilon(X,Z_0)$ is determined as follows. For a given x, the probability that $\mathrm{sgn}(Y) \neq \mathrm{sign}\ (x)$ is

$$\begin{aligned} \epsilon(x) &= \frac{1}{2}\Pr\{Z_0^2 > x^2\} + \frac{1}{4}\Pr\{Z_0^2 = x^2\} \qquad & x \neq 0 \\ &= \frac{1}{2} & x = 0. \end{aligned}$$

Since Z_0^2 has a continuous density for $z \neq 0$, $\Pr\{Z_0^2 = x\} = 0$, unless $x = 0$. Thus the transition probability $\epsilon(X,Z_0)$ is given by

$$\varepsilon = \frac{1}{2}\Pr\{X \neq 0\} + \frac{1}{2}E(\Pr(Z_0^2 > X^2))$$

Now from (4.11), $\Pr\{Z_0^2 > x^2\} = \max\{\frac{1}{A}(1 - \frac{x^2}{2A}),0\}$, and so if $\rho = \Pr\{X \neq 0\}$,

$$\epsilon \geq \frac{1}{2}(1-\rho) + \frac{1}{A}\left(1 - \frac{E(X^2)}{2A}\right)$$

$$\geq \frac{1}{2}(1-\rho) + \frac{1}{4A}$$

$$\geq \frac{1}{4A}.$$

Hence $\epsilon(X,Z_0) \geq \frac{1}{4A}$, provided only $E(X^2) \leq A$. It follows that

$$\phi(X,Z_0) \leq \log 2 - H(1/4A)$$

$$= \phi(X_0,Z_0).$$

To prove the other half of (4.16), let X_0 be as described in (4.11), and let Z satisfy $E(Z^2) \leq 1$. Since $\phi(X_0,Z)$ is convex $\cup$ in Z ([8], Theorem 1.7), we may assume Z is symmetric, in which case, again, $\mathrm{sgn}(X_0)$ and $\mathrm{sgn}(Y)$ are connected by a BSC. For a fixed z, the probability of error is

$$\epsilon(z) = \frac{1}{2}\Pr\{X_0^2 < z^2\} + \frac{1}{4}\Pr\{X_0^2 = z^2\}, \qquad z \neq 0$$

$$= \frac{1}{2}\Pr\{X_0 = 0\} \qquad , \qquad z = 0.$$

But X_0^2 is uniform on $[0,2A]$, and so $\Pr\{X_0^2 = z^2\} = 0$ for all z. Thus for all z

$$\epsilon(z) = \frac{1}{2}\Pr\{X_0^2 < z^2\}$$

$$= \frac{1}{2}\min\{\frac{z^2}{2A}, 1\}, \text{ from (4.11)}.$$

Hence the transition probability $\epsilon = E(\epsilon(Z))$ satisfies

$$\epsilon \le \frac{1}{2} \cdot \frac{E(Z^2)}{2A} \le \frac{1}{4A}.$$

Hence $\epsilon(X_0,Z) \le 1/(4A)$. This establishes the saddlepoint value of ϵ as $1/(4A)$ (for $A \ge 1$), and so also the saddlepoint value of ϕ as $\phi(X_0,Z_0) = \log 2 - H(1/4A)$.

Proof of Theorem 4.3. Choose any $A > 0$, and let X_0 and Z_0 be as described in the statement of Theorem 4.3. We begin by showing that

$$\phi(X,Z_0) \le \phi(X_0,Z_0)$$

for any X satisfying $E(X^2) \le A$. The payoff function can be written as

$$\phi(X,Z_0) = I(X;Y) = H(Y) - H(Y|X),$$

where the expected conditional entropy $H(Y|X)$ is the expectation with respect to X of the quantity*

$$H(Y|X = x) = H(P(Y = -1|X = x)). \tag{4.18}$$

But

$$P(Y = -1|X = x) = P(X + Z_0 < 0|X = x) + \frac{1}{2}P(X + Z_0 = 0|X = x)$$

$$= P(Z_0 < -x) + \frac{1}{2}P(Z_0 = x).$$

*Throughout this proof we will use the letter H in two different function contexts, as the entropy operator or random variable, and as the entropy function of a real variable: $H(x) = -x\log x - (1-x)\log(1-x)$.

So

$$P(Y = -1 | X = x) = \begin{cases} G_0(-x) & \text{for} \quad x \neq 0 \\ \frac{1}{2} & \text{for} \quad x = 0. \end{cases} \tag{4.19}$$

It follows from (4.18), (4.19) and (4.14) that

$$H(Y | X = x) = \begin{cases} -\lambda x^2 + \mu & \text{for} \quad x = 0 \\ \log 2 & \text{for} \quad x = 0 \end{cases} \tag{4.20}$$

Since $\mu \leq \log 2$, we have (denoting the distribution of X by F) that

$$\begin{aligned} H(Y|X) &\geq \int (-\lambda x^2 + \mu) dF(x) \\ &\geq -\lambda A + \mu. \end{aligned} \tag{4.21}$$

Also observe that since $Y = \text{sgn}(X + Z_0)$ assumes only two values, we have

$$H(Y) \leq \log 2 \tag{4.22}$$

Hence we have in combination with (4.21) that

$$\phi(X, Z_0) \leq \log 2 + \lambda A - \mu, \tag{4.23}$$

with equality if and only if the distribution of X is symmetric about (4.24a) zero, which is the condition for equality in (4.22); and

$$P\{X = 0\} = 0 \text{ if } \mu > \log 2, \ P(|X| \geq \sqrt{\mu/\lambda} = 0, \text{ and } E(X^2) = A, \tag{4.24b}$$

which are the conditions for equality in (4.21).

The random variable X_0 has all these properties. Thus we have shown that

$$\phi(X,Z_0) \le \phi(X_0,Z_0) \tag{4.25}$$

for all random variables satisfying $E(X^2) \le A$. This completes the first half of the proof of optimality of the pair (X_0,Z_0). (Elementary calculations show that $\phi(X_0,Z_0) = C^*(A)$.)

To finish the proof of optimality, we need to establish that

$$\phi(X_0,Z_0) \le \phi(X_0,Z) \tag{4.26}$$

for all Z satisfying $E(Z^2) \le 1$.

We indicate the dependence of $\phi(X_0,Z)$ on the distribution G (say) of Z by writing

$$\begin{aligned}\omega(G) &= \phi(X_0,Z) = H(Y) - H(Y|X_o)\\ &= H(\int G^*(-x)dF_0(x)) - \int H(G^*(-x)dF_0(x),\end{aligned}$$

where $G^*(s) = \frac{1}{2}[G(x) + G(x-)]$. We will show that

$$\phi(X_0,Z_0) = \min\{\omega(G_0 + \eta) : \eta \in S\}, \tag{4.27}$$

where S denotes the class of all real valued functions of bounded variation on $(-\infty,\infty)$ such that

$$d\eta(x) = 0, \tag{4.28a}$$

$$x^2 d\eta(x) \le 0, \text{ and} \tag{4.28b}$$

$$0 \le G_0(x) + \eta(x) \le 1 \text{ for } -\infty < x < \infty. \tag{4.28c}$$

Observe that S is convex. Any η satisfying these conditions will be called admissible. Once (4.28) is proven we will have (4.26) since $\eta = G - G_0$ is in S.

Note that the admissible η's may be assumed to have support in $(-\sqrt{\mu/\lambda}, \sqrt{\mu,\lambda})$. Consider a distribution function $G = G_0 + \eta$, where support does not lie entirely in $(-\sqrt{\mu/\lambda}, \sqrt{\mu/\lambda}$.

$$G_1(x) = \begin{cases} G(x) & \text{for } x \in (-\sqrt{\mu/\lambda}, \sqrt{\mu/\lambda}) \\ 0 & \text{for } x \le -\sqrt{\mu/\lambda} \\ 1 & \text{for } x \ge \sqrt{\mu/\lambda} \end{cases}$$

Since $F_0'(x) = 0$ when $|x| > \sqrt{\mu/\lambda}$, it is easy to see that $\omega(G) = \omega(G_1)$. Moreover, we can write $G_1 = G_0 + \eta_1$, where η_1 has support in $(-\sqrt{\mu/\lambda}, \sqrt{\mu/\lambda})$.

The Gateaux derivative of ω at G_0 in any admissible direction η, $D_\eta\omega(G_0)$, splits naturally into the difference of two terms. The first term is

$$H'\left(\int G_o^*(-x)dF_0(x)\right) \cdot \int \eta(-x)dF_0(x).$$

However, the argument of H' is $P(Y = -1) = \frac{1}{2}$ (by symmetry of G_0 about zero) and thus this term vanishes. Therefore,

$$D_\eta\omega(G_0) = -\int H'(G_0^*(-x)\eta(-x)dF_0(x).$$

By the remarks about F_0' just given, this integral can be restricted to the interval $[-\sqrt{\mu/\lambda}, \sqrt{\mu/\lambda}]$. Also note that by symmetry of G_0 for $x \neq 0$ $G_0(-x) = 1 - G_0(x)$ so that $G_0'(-x) = G_0'(x)$. Hence by the properties of G_0 and G_o' given in the statement of Theorem 4.3 for $x \neq 0$

$$-2\lambda x = \frac{d}{dx} H(G_0^*(-x)) = -H'(G_0(-x))G_o'(-x)$$

$$= H'(G_0(-x)G_o'(x) = -\frac{1}{A} H'(G_o^*(-x)F_0'(x).$$

Thus for $x \neq 0$

$$H'(G_o^*(-x)dF_0(x) = 2A\lambda x\ dx.$$

On the other hand when $x = 0$, either $G_o^*(-x) = \frac{1}{2}$ or $F_0(x)$ is continuously differentiable, so that the integral for $D_\eta\omega(G_0)$ receives no contribution at $x = 0$. We have thus deduced that

$$D_\eta\omega(G_0) = -\int_{\sqrt{\mu/\lambda}}^{\sqrt{\mu/\lambda}} \eta(-x)2A\lambda x\ dx.$$

Applying integration by parts and (4.28a) we see that this last expression equals

$$-A\lambda \int_{-\sqrt{\mu/\lambda}}^{\sqrt{\mu/\lambda}} y^2\ d\eta(y),$$

which by the fact that $-A\lambda \leq 0$ along with (4.28b) is nonnegative. This proves that G_0 is a local minimizer of ω. However, since ω is a convex function of $G = G_0 + \eta$, G_0 is in fact a global minimum of ω on A. This completes the proof of Theorem 4.3.

5. Spread-Spectrum Communication: The Results of Houston and Viterbi-Jacobs.

We conclude this article by placing our previous results in a larger setting - the setting of spread-spectrum communications.

During World War II, researchers in telecommunications, motivated by the urgent need to communicate rapidly and reliably under harsh battlefield conditions, and fueled by almost unlimited government funding, made prodigious advances. Two of the most important outcomes of this wartime research were Claude Shannon's invention of Information Theory, and the development of Spread-Spectrum Communication techniques.

It is not possible to identify a unique "father" of Spread-Spectrum, bearing the same relationship to SS that Shannon does to Information Theory. (In a recent survey article Scholtz [14] traces the gradual emergence of a coherent SS discipline from a confusing tangle of post-World War II hardware developments.) Nevertheless, there is a unifying insight, which might be stated as follows.

The Philosophy of Spread-Spectrum

> "Suppose we must transmit a d-dimensional signal in the presence of a jammer with finite power. If this signal is disguised as an n dimensional signal, $n > d$, then the jammer will be forced to distribute his power over n dimensions, and his power will be effectively reduced by the factor n/d."

There are of course many details to be worked out before this philosophy can be turned to practical use.

The first detail is the question of how to disguise, i.e. to spread the spectrum of, the transmitted signal. There are two major classes of techniques in widespread use. The first technique is called direct-sequence spreading. In it the transmitted signal is multiplied by a binary pseudorandom sequence whose symbol rate (chip rate) is many times the data rate (bit rate). The spreading factor n/d, usually called the processing gain, for a direct sequence system equals the ratio of the chip rate to the bit rate. Pursley [13] has written a very good survey article on the issues involved in direct-sequence spreading.

The second major kind of SS technique is frequency hopping. Here the signal is transmitted in "unspread" form, except that periodically the carrier frequency is "hopped" to another location. The hopping pattern is pseudorandom. If there are N different carrier frequencies used (and if they are at least W Hz apart, if W is the data bandwidth), the processing gain for the FH system is N. The recent survey article by Pickholtz et al. [12] has a good coverage of FH techniques.

In the overall theory of SS techniques, the spreading technique is very important. However, we shall assume the processing gain is given, and focus on another issue, which we now describe. (See also Viterbi [17].)

Suppose the data rate is R bits/second, so that the original unspread bandwidth is 1/R Hz. If the spread bandwidth is W Hz., then the processing gain is given by

$$\text{P.G.} = W/R. \tag{5.1}$$

If the jammer's total power is J watts, the average jamming noise density will be

$$N_0 = J/W \quad \text{watts/Hz}. \tag{5.2}$$

If the received signal power is S watts, the received energy per bit is

$$E_b = S/R \quad \text{joules}. \tag{5.3}$$

Combining (5.2) and (5.3), we have the important relationship

$$\frac{J}{S} = \frac{W/R}{E_b/N_o} \tag{5.4}$$

The quantity J/S appearing in (5.4) is known as the jamming margin. It represents the tolerable jamming power-to-signal power ratio. In antijam systems, it is obviously desirable to have as large a jamming margin as possible. From (5.4), we see that this can be accomplished in two ways. We can either increase the numerator, which is the processing gain (see (5.1)), or we can decrease the denominator E_b/N_o. It is the latter possibility that is our chief concern here. It turns out that for a given processing gain, the difference in jamming margin (vs. a sophisticated jammer) between a sophisticated AJ system and an unsophisticated one can be 30-40 dB, or more.

Until 1975, almost all of the published literature on SS/AJ dealt with techniques for improving the processing gain. Then, almost simultaneously, there appeared two key papers which dealt with the important E_b/N_0 issue. These were Houston [7] and Viterbi and Jacobs [18]. Let us now briefly describe these important results.

What Houston showed is that the jammer is not as vulnerable to increasing processing gains as Eq. (5.4) might indicate, and that as W/R increases, he can adopt jamming strategies that make E_b/N_0 increase too. Houston considered several different kinds of SS systems, but we can illustrate his basic results with what is now a classic example, FH/binary FSK vs. an optimal partial-band jammer. (Sec. 2.1 of [7].)

It is well-known that if binary FSK modulation is used in the presence of wideband white Gaussian noise, the detected bit error probability ε is given by the formula

$$\varepsilon = \frac{1}{2} e^{-\frac{1}{2}x} \quad , \quad x = E_b/N_o, \tag{5.5}$$

where E_b is the received energy per bit (as in (5.3)), and N_0 is the spectral density of the noise. (Reference [20], Eq. (7.68). If we invert (5.5), we can write

$$E_b/N_0 = 2 \log(2\varepsilon), \tag{5.6}$$

which gives the needed E_b/N_o as a function of the desired bit error probability ε. Now suppose that this noise is due to a jammer, in a FH/SS system. We assume that the jammer cannot know the communicators hopping pattern, which appears to him only as a uniform random distribution among the N possible carrier frequencies. If the jammer decides to distribute his power uniformly over the entire spread-spectrum bandwidth W, then Eq. (5.6) continues to describe the relationship between E_b/N_0 and ε. Houston observed, however, that the jammer can do much better than this. He can, for example, decide to distribute his power uniformly over only a small fraction of the total spread bandwidth. If this fraction is denoted by ρ, $0 \le \rho \le 1$, then his noise density will be increased to N_0/ρ in the fraction of the band occupied, and will of course be 0 in the unoccupied portion of the band. The resulting bit error probability ε will then be $\frac{1}{2} e^{-x/2\rho}$ with probability ρ, and zero with probability $1-\rho$. Thus

$$\varepsilon = \rho \cdot \frac{1}{2} e^{-x/2\rho} \tag{5.7}$$

Notice that Eq. (5.7) reduces to Eq. (5.5), with $\rho = 1$. The jammer will naturally wish to choose ρ so as to minimize ε. This is an easy exercise in calculus. If we denote the minimum of (5.7) over all $0 \le \rho \le 1$ by ε^*, the result is

$$\varepsilon^* = \begin{cases} \frac{1}{2} e^{-x/2} & \text{if } x \le 2 \quad (\rho = 1) \\ \frac{1}{ex} & \text{if } x \ge 2 \quad (\rho = \frac{2}{x}). \end{cases} \tag{5.8}$$

or, inverting (5.8), we get

$$E_b/N_o = \begin{cases} -2\log(2\varepsilon), & \text{if } \varepsilon \ge \frac{1}{2e} = .1845 \\ \frac{1}{e\varepsilon}, & \text{if } \varepsilon \le .1845. \end{cases} \tag{5.9}$$

The difference between (5.6) and (5.9) is dramatic. For example, if $\varepsilon = 10^{-5}$, a commonly-cited requirement in telecommunications systems, (5.6) says that a uniform jammer requires E_b/N_o of 21.64 (13.35 dB), whereas (5.8) shows that a worst-case partial band jammer forces E_b/N_o up to 36788 (45·7 dB). The difference is a factor of 1700 (32 dB). In other words, the jamming margin is reduced 32 dB if the jammer changes from a cooperative uniform jammer to a sophisticated partial band jammer.

Although this was a rather specific example, it typifies Houston's more general results: an optimized partial-band, or tone jammer can wreak havoc on an unsophisticated SS system, and the simplified analysis leading to (5.4) can be very misleading. Since Houston's pioneering work, many other researchers have reached similar conclusions about other specific SS systems. (Theorem 4.2, above, is another example. There we saw that the BSC crossover probability ε^* was also an inverse-linear function of the signal-to-noise ratio.)

In summary, Houston showed that brute-force spectrum spreading does not by itself guarantee adequate AJ capability, if the jammer is clever. But two years later, in 1977, Viterbi and Jacobs [18] showed that spectrum-spreading combined with error-corrector coding, can neutralize a wide variety of sophisticated jamming threats. More recently, we have developed a different approach [9] [10] which yields the same conclusion. We conclude this section with a description of these important results. Again we illustrate using the FH/SS-FSK system to illustrate.*

Consider first an unsophisticated jammer who distributes his noise power uniformly over the entire spread-spectrum bandwidth. As we saw in Eq. (5.5) above, this results in an error probability of $\varepsilon = \frac{1}{2} e^{-1/2x}$, where x is the bit signal-to-noise ratio. Effectively, then the communicator is presented with a binary symmetric channel, as shown in Figure 7.

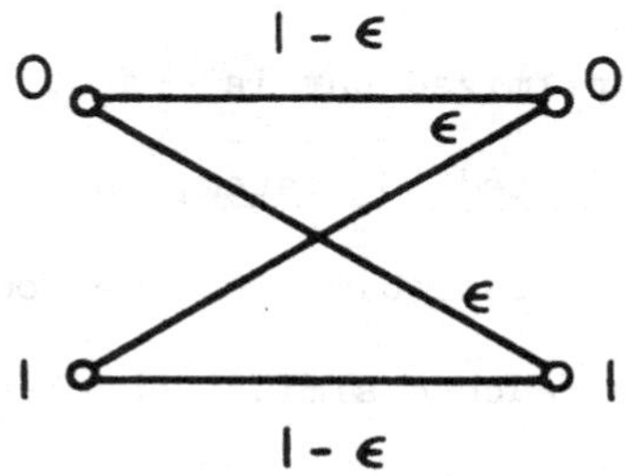

Figure 7. A binary symmetric channel.

Now it is well-known that the (Shannon) capacity of this channel is $C = 1-H_2(\varepsilon)$, where $H_2(\varepsilon) = -\varepsilon\log_2\varepsilon-(1-\varepsilon)\log_2(1-\varepsilon)$ is the binary entropy function. In the presence of the uniform noise jammer, this means that reliable coded communication is in principle possible, provided that the dimensionless code rate is no larger than $1-H_2(\varepsilon)$, where ε is given by (5.5). Since x measures the per letter signal-to-noise ratio, and r measures the number of information bits per letter, it follows that the bit signal-to-noise ratio E_b/N_o is given

*This example is our own [9], [10]. Viterbi's work uses the parameter R_o as the payoff function rather than channel capacity. We feel this is the simplest way to make Viterbi's point.

$$E_b/N_o = \frac{x}{r} \geq \frac{x}{1-H_2(\varepsilon)} \cdot \tag{5.10}$$

The right-hand side of (5.10) is a function of the per-letter SNR x; which can be controlled by the communicator by varying the code rate r. In view of formula (5.4) for the jamming margin, the communicator will naturally wish to minimize E_b/N_o. The minimum value of the function $x/(1-H_2(\varepsilon))$ which appears on the right side of (5.10) occurs at x = 3.05, corresponding to a channel crossover probability of $\varepsilon = .109$, and a code rate of r = .503. Thus in principle, the minimum E_b/N_o required for reliable <u>coded</u> communication is from (10) 3.05/.5034 = 6.06. In "dB", this is 7.82:

$$(E_b/N_o)_{min} = 7.82 \text{ dB} \qquad \text{(uniform jammer)} \tag{5.11}$$

This is a significant gain over the required E_b/N_o for $P_{bit} = 10^{-5}$ for an <u>uncoded</u> system, which we computed above to be 13.3 dB. But a much more surprising result occurs when we consider coding vs. an <u>optimized partial-band jammer</u>.

For the PB jammer, we can repeat the previous argument exactly, except that we must replace formula (3.5) with formula (5.8), which applies to the PB jammer. In this case we find, after some computation, that the minimum of $x/(1-H_2(\varepsilon))$ occurs at x = 2.38, $1-H_2(\varepsilon) = .3790$. Hence the minimum required E_b/N_o of a coded FH-SS system vs. a PB jammer is 2.38/.3790 = 6.28 = 7.98dB:

$$(E_b/N_o)_{min} = 7.98 \text{ dB} \qquad (5.12)$$

(partial-band jammer)

This is very surprising. First of all it is a huge gain (45.7 - 8.0 = 37.7dB) over the performance of an uncoded SS system vs. partial-band jamming. But even more surprising, the minimum coded E_b/N_o required vs. a p.b. jammer is only very slightly more (8.0-7.8 = 0.2dB) than the minimum required vs. an unsophisticated uniform jammer. These numbers are displayed in Table 1:

		communicator: uncoded	communicator: coded	
jammer	uniform	13.3	7.8	5.5
jammer	partial-band	45.7	8.0	37.7
	differences	32.4	0.2	

Table 1. E_b/N_o required for $P_b = 10^{-5}$.

Thus if sophisticated coding is used, the advantage to the jammer of adopting a PB strategy disappears. In fact, the potential coded performance vs. a PB jammer is 5dB superior to the uncoded performance vs. a uniform jammer!

Although we have considered in detail only one special case, this conclusion actually holds quite generally, and explains why coding is (or ought to be) so widespread in actual SS systems. This insight is so important that we state it as an addendum to the philosophy of SS given earlier:

The Philosophy of Spread-Spectrum (Addendum):

> "A sophisticated jammer might not choose to distribute his power uniformly over all n dimensions, in which case the previous argument is misleading, and in fact grossly inadequate signal protection may result. However, the proper use of error-correction coding can effectively homogenize a non-uniform jammer, and produce system performance as good as or better than it would be vs. a benign, uniform jammer."

6. Acknowledgements

The results in this paper have been developed over a period of several years, and some of them have already appeared in print, or will appear in print shortly. The abstract Communicator vs. Jammer game-theoretic formulation presented in Section 2 of this paper was first published by Wayne Stark and me in [9]. (It also appears in the early papers of Blackman [3] and Dobrushyn [6], but was apparently not pursued by later researchers, until our own work.) Stark's Ph.D. thesis [16] contains much more along these lines. The pseudorandom scrambler described in Section 3 is described in more detail in the important forthcoming paper on interleaving by Berlekamp and Tong [2]. The Gaussian saddlepoint theorem of Section 4 (Theorem 4.1) is not new. It is stated by both Blachman [3] and Dobrushyn [6] in the papers already mentioned, but its proof depends so heavily on inequalities proved by Shannon in his original paper on information theory, that I have attributed it to all three of these authors. Theorem 4.2 appears in print for the first time here, but an extension of these ideas will be found in a forthcoming paper by Eugene Rodemich and me [11]. Theorem 4.3 is due to Martin Borden, David Mason, and me, and will appear in our forthcoming paper [11]. As mentioned in the text, the results of Section 5 are essentially due to Houston [7], and to Viterbi and Jacobs [18]. (Viterbi and Jacobs measure the effectiveness of coding with the cutoff parameter R_0 rather than channel capacity C, but they were the first to see the importance of the 'addendum' quoted at the end of Section 5.)

A portion of this work was supported by the Joint Services Electronics Program under Contract N00014-79-C-0424, while I was employed by the University of Illinois.

References

1. Barbu, V. and V. Precupanu, Convexity and Optimization in Banach Spaces. Bucharest: Editura Academiei, 1978.

2. Berlekamp, E. R., and Po Tong, "Interleaving," IEEE Trans. Inform. Theory, in press.

3. Blachman, N. M., "Communication as a Game," 1957 WESCON Conference Record, pp. 61-66.

4. Blachman, N. M., "The Convolution Inequality for Entropy Powers," IEEE Trans. Inform. Theory, vol. IT-11 (1965), pp. 267-271.

5. Borden, J. M., D. M. Mason, and R. J. McEliece, "Some Information-Theoretic Saddlepoints," SIAM J. Opt. Control, in press.

6. Dobrushyn, R. L., "Optimal Information Transmission over a Channel with Unknown Parameters," Radiotekh. Elektron., vol. 4, no. 12, (1959), pp. 1951-1956.

7. Houston, S. W., "Modulation Techniques for Communication. Part 1: Tone and Noise Jamming Performance of Spread-Spectrum M-ary FSK and 2,4-ary DPSK Waveforms," Proc. 1975 IEEE National Aeronautics and Electronics Conference (NAECON), pp. 51-58

8. McEliece, R. J., The Theory of Information and Coding. Reading, Mass.: Addison-Wesley, 1977.

9. McEliece, R. J. and W. E. Stark, "An Information-Theoretic Study of Communication in the Presence of Jamming," Proc. 1981 IEEE International Conference on Communications, pp. 45.3.1-45.3.5.

10. McEliece, R. J. and W. E. Stark, "The Optimal Code Rate vs. a Partial-Band Jammer," Proc. 1982 Military Communications Conference, pp. 8.5.1-8.5.4.

11. McEliece, R. J. and E. R. Rodemich, "Communication with MFSK vs. Optimal Jamming," paper in preparation.

12. Pickholz, R. L., D. L. Schilling, and L. B. Milstein, "Theory of Spread-Spectrum Communication - a Tutorial," IEEE Trans. Comm., vol. COM-30 (1982), pp. 855-884.

13. Pursley, M. B., "Spread-Spectrum Multiple-Access Communication," pp. 139-199 in Multi-User Communication Systems, G. Longo, Ed., New York: Springer-Verlag, 1981.

14. Scholz, R. A., "The Origins of Spread-Spectrum Communication," IEEE Trans. Comm., vol. COM-30 (1982), pp. 822-854.

15. Shannon, C. E., "A Mathematical Theory of Communication," Bell Syst. Tech. J. 27 (1948) pp. 379-423, 123-656.

16. Stark, W. E., Coding for Frequency-Hopped Spread-Spectrum Channels with Partial-Band Interference, Ph.D. thesis, University of Illinois at Urbana-Champaign, 1982. (Reprinted as U. of I. Technical Report no. UILU-ENG 82-2211, July 1982).

17. Viterbi, A. J., "Spread-Spectrum Communications - Myths and Realities," IEEE Comm. Mag., May 1979, pp. 11-18.

18. Viterbi, A. J. and I. M. Jacobs, "Advances in Coding and Modulation for Noncoherent Channels Affected by Fading, Partial-Band, and Multiple-Access Interference," pp. 279-308 in Advances in Communication Systems, vol. 4, New York: Academic Press, 1975.

19. Wolfowitz, J., Coding Theorems of Information Theory, 3rd Ed., Berlin: Springer Verlag, 1978.

20. Wozencraft, J. M. and I. M. Jacobs, Principles of Communication Engineering, New York: Wiley, 1965.

CONFLICT RESOLUTION PROTOCOLS FOR SECURE MULTIPLE-ACCESS COMMUNICATION SYSTEMS

Toby Berger and Nader Mehravari
School of Electrical Engineering
Cornell University
Ithaca, NY 14853 USA

Significant demand exists for systems which permit a large population of occasionally active communicators to exchange messages securely. In these lectures we introduce and analyze efficient algorithms for resolving conflicts over access to the transmission medium among communicators who are employing either public key cryptography or spread spectrum signaling. Our intention is to extend some recent results in the theory of random multiple-access time-slotted communication systems to make them compatible with constraints imposed by the need for secure communications.

LECTURE 1
INTRODUCTION

1.1 Background and Motivation

Designing a cost-effective communication system necessitates efficient utilization of the communication links and processing equipment. Resource sharing is a key idea in achieving this goal. Communication engineers have long made use of sophisticated resource sharing techniques in order to achieve efficient use of the available communication resources. Resource sharing, or multiple accessing, provides the means by which many senders and receivers of information share a common communication link. The traditional solution to the multiple accessing (MA) problem has been channel-oriented. In a channel-oriented, or circuit-switched MA system, the total transmission capacity available is subdivided into separate channels in space, time, or frequency. Each of the several subscribers is given the sole access to one of the subchannels, and there is no interference between the subscribers. Time division multiple accessing (TDMA) and frequency division multiple accessing (FDMA) are familiar examples of channel-oriented MA systems. (For a complete categorization of MA modulation techniques see [1] and [2].)

The channel-oriented MA schemes have proven their worth in a number of applications involving voice traffic and data traffic. However, some communicators operate in a "bursty" nature. A "bursty" source of information has nothing to send most of the time; i.e., the ratio between the peak and average data rate is high. (For a quantitative measure of burstiness, see [3].) Hence, in a channel-oriented MA system, a portion of total transmission capacity is wasted when a user is granted sole access to the communication link but has nothing to send. This makes the class of channel-oriented MA systems unattractive for situations with a large population of bursty users.

Another solution to the MA problem is random multiple accessing, which is also referred to as random accessing (RA). In a RA system every user is allowed to occupy the communication link when he has information to transmit. However, since there is no coordination among the users, it may happen that messages from different users interfere with one another. In that case a so-called confusion resolution algorithm (CRA) is used by the communicators to control the retransmission of messages that collided. The difference between channel-oriented MA schemes and RA schemes is analogous to the difference between the way traffic is handled at a busy downtown intersection and a lightly traveled one. Experience has shown that a traffic light is necessary at a heavily used downtown intersection. However, at a lightly used intersection it suffices to have a stop sign.

Today's computer technology has generated many MA situations characterized by a large population of users, all wanting to communicate on a common link, but each of whom has nothing to say most of the time. Computer communication networks, local area networks, and radio broadcast systems are examples of applications in which RA systems are useful. The benefits to be derived from incorporating a secure communication capability into these and other applications have been cited repeatedly and therefore need not be reiterated here. The combination of these considerations has motivated our analysis of RA systems that are compatible with public key or spread spectrum secure communication schemes.

1.2 Random-Accessing Problem

Consider a population of spatially isolated, bursty, independent users trying to communicate either with a central facility or with each other on a common channel. The sources generate messages of fixed length, called packets. The generation times of the packets are independent and form a point process. The channel is time-slotted; that is, the channel time is divided into equal segments called slots. Slots are numbered k=1,2,3,.... The length of each slot equals that of a packet. It is assumed that the sources are synchronized with the channel in the sense that the attempts to transmit packets are exactly time-aligned with the slots. A source may generate a packet at any time, but it may transmit it only during the slots specified by the random-accessing algorithm (RAA) being used.

Whenever two or more sources try to use the same slot, a collision occurs. In that event none of the messages reaches its destination and all require retransmission. We further assume that SNR is high enough, and/or enough error correction bits are applied to each packet so that, if only one message is transmitted in a given slot, it reaches its destination error free. Finally, we assume that at the end of the kth slot, each source learns the value of a d-ary feedback random variable $Z_d(k)$ that provides some information about the number of packets that were transmitted therein. Each user knows the history of his own transmissions but learns about the transmission history of others only via the feedback. We will call such a channel a "random multiple-access, collision-type, time-slotted, packet-switched communication channel with d-ary feedback." For the sake of simplicity we shall henceforth refer to it as a random multiple-access channel, or simply a random-access channel.

DEFINITION 1.2.1 - A random accessing algorithm (RAA) is a protocol that the users follow to access a random multiple-access channel.

DEFINITION 1.2.2 - A confusion resolution algorithm (CRA) is a protocol for transmission and retransmission of packets by individual users with the property that, after each collision, all packets involved in the collision are eventually retransmitted successfully. Also, all users (not only those whose packets collided) eventually and simultaneously become aware of when these packets have been successfully retransmitted. (This definition is due to Massey [18].)

We say that a collision is resolved when the transmitters become aware that all the collided packets have been successfully retransmitted. A CRA is the main part of a RAA. To obtain a RAA from a CRA, it suffices to specify the rule by which a user with a new packet will determine the slot for its initial transmission. Thereafter, the users follow the CRA (if necessary) to determine when the packets should be retransmitted.

The objective in designing RAAs and CRAs is to maximize the fraction of slots devoted to exactly one packet (i.e., neither empty nor wasted because of a conflict). We will use different performance measures depending on whether we are dealing with an infinite or a finite user model. The performance measures for finite and infinite user models will be defined in Lectures 2 and 3, respectively.

1.3 Classification of Random Multiple-Accessing Problems According to the Available Feedback

In a random multiple-access channel, at the end of the kth slot each user learns the value of a d-ary feedback random variable, $Z_d(k)$. We classify the random multiple-accessing problems according to the number, d, of different possible values of the feedback random variable. We divide the class of random multiple-accessing problems into three subclasses according to whether d=2, or d=3, or d>3.

1.3.1 (0,1,e) - Ternary Feedback, d=3. This is the first and the most thoroughly studied type of feedback for a RA system. Here, $Z_3(k)$, the ternary feedback is defined as

$$Z_3(k) = \begin{cases} 0 & \text{if no packets in slot } k \\ 1 & \text{if exactly one packet in slot } k \\ e & \text{if} \geq 2 \text{ packets in slot } k \end{cases}$$

This type of feedback, known as 0,1,e-feedback, was first introduced in the ALOHA system. Algorithms for this feedback will be reviewed in Section 1.4.

1.3.2 (0,1,2,...,d-2,e) d-ary Feedback, d>3. This d-ary feedback random variable, d>3, is defined as:

$$Z_d(k) = \begin{cases} i & \text{if i packets in slot k, 1=0,1,2,...d-2} \\ e & \text{if} \geq \text{d-2 packets in slot k} \end{cases}$$

This type of feedback was first introduced by Tsybakov [28] for studying resolution of conflicts of known multiplicity. However, he only considered the special case where d = ∞. Georgiadis and Papantoni-Kazakos [29] studied this type of feedback for general d in a model for a RA system with energy detectors.

1.3.3 Binary Feedback, d=2. Binary feedback is the main topic of these lectures. We introduce three different types of binary feedback:

$$Z_{CNC}(k) = \begin{cases} C & \text{if} \geq \text{2 packets in slot k (conflict)} \\ NC & \text{if 1} \leq \text{packet in slot k (no conflict)} \end{cases}$$

$$Z_{SN}(k) = \begin{cases} S & \text{if} \geq \text{1 packet in slot k (something)} \\ N & \text{if no packets in slot k (nothing)} \end{cases}$$

$$Z_{SF}(k) = \begin{cases} S & \text{if one packet in slot k (success)} \\ F & \text{if either no or} \geq \text{2 packets in slot k (failure)} \end{cases}$$

The conflict/no conflict (CNC) feedback informs the users only about whether or not there was a conflict in the previous slot. The something/nothing (SN) feedback informs the users about whether or not the previous slot was empty. This type of feedback is characteristic of public-key secure computer communications. If, for example, user i is the only user who has transmitted a packet during the kth slot, then simply by listening to the channel he will be aware of the fact that the packet has been successfully transmitted. Meanwhile, any user j≠i other than the intended recipient cannot distinguish between a single message encrypted for someone else and a collision between two or more messages. This situation results in SN feedback. The success/failure (SF) feedback informs the users whether or not the previous slot contained exactly one message. This situation arises, for example, if the receivers cannot distinguish between channel noise and collision noise. Spread spectrum RA systems result in SF-like feedback in the event that the transmitted power is kept low enough that the noiselike waveform which results from the collision of two or more transmitted signals is difficult to distinguish reliably from noise alone. CNC binary feedback does not seem to have any readily apparent applicability to secure communication systems, so it will not be included in the sequel.

1.4 Review of Existing Random Accessing Algorithms

We will briefly discuss a number of schemes previously introduced and analyzed for the infinite-user model with 0,1,e-ternary feedback. Table 1.4.1 provides a capsule guide to the literature.

	Confusion Resolution Algorithms	Bounds on the Performance of the Optimum Protocol
Binary Feedback	*, [11], [30], [31]	*
Ternary Feedback (0,1,e)	[4-20]	[21-27]
d-ary Feedback d>3	[28], [29]	[21]

[ ] - references containing the results

* - studied herein

TABLE 1.4.1 - Summary of our results and thier relation to previous work.

Consider a random-access channel with a large, effectively infinite, population of sources. The generation times of the packets by the users is modeled as a Poisson point process on the time interval $T = (a,b]$ with intensity $\lambda>0$. Therefore, we can associate with each packet the time it was generated. We assume that users are able to listen to the channel and determine instantaneously whether 0, 1, or more than one message was sent. Alternatively, this feedback can be provided by a central facility through a feedback channel. Efficiency (throughput) is defined as the fraction of slots on the channel devoted to exactly one packet.

1.4.1 ALOHA and its Modifications. The first formal RA system was introduced by Abramson [4]. This was a system for allowing remote terminals on various islands of Hawaii to communicate with a central computer via a common radio broadcast channel. This system was not a time-slotted one, i.e., information packets were not time aligned with the channel slots. This system is now known as pure-ALOHA. in pure-ALOHA, as soon as a user generates a new packet, he transmits it on the common channel. When a source determines that his packet has collided, he retransmits that packet after a randomly chosen time. Assuming a Poisson traffic for the newly generated packets, Abramson showed that pure-ALOHA has a maximum throughput of $1/2e = 0.1839$ under the approximation that the total channel traffic (new plus retransmissed traffic) is also Poisson. It was noticed by Roberts [5] that the throughput could be increased to $1/e = 0.3679$ by using a time-slotted channel. This modification of pure-ALOHA is known as slotted-ALOHA.

In order to analyze pure-ALOHA, Abramson introduced an assumption of "statistical equilibrium" which implied that the retransmission traffic forms a Poisson process independent of new packet traffic. However, it can be shown that, instead of reaching a "statistical equilibrium" such a system may become unstable. That is, it may reach a state where many sources are attempting to get access to the channel, and very few are succeeding [6-9]. However, unstability can be avoided by using some auxilary control [7-11].

1.4.2 The Tree Algorithm and its Modifications. A binary tree search algorithm for scheduling retransmission of packets that collided was first treated by Capetanakis [12,13,14], and by Tysbakov and Mihailov [15]. Stability is among the desired properties of this protocol. In this scheme each user chooses an address which specifies his location on a binary tree. The addresses are chosen either deterministically or randomly, say by flipping a coin. When a collision occurs, the tree is divided in half and sources in the top half attempt retransmission first. The sources in the bottom half must wait until all conflicts in the top half have been resolved before they may attempt retransmission. If, upon division, further conflicts occur, the process is repeated. We will refer to this protocol as the tree-CRA. The time period between the slot in which a collision first occurred and that in which it is finally resolved is called a confusion resolution interval (CRI). In order to use this tree-CRA as an RAA, it suffices to specify the rule by which a user with a new packet will determine the slot for its initial transmission. One obvious such rule is to transmit a new packet in the first slot following the CRI in progress when the packet was generated. This RAA was analyzed for the Poisson infinite user model [12,13], and was shown to have a throughput of 0.346 packets per slot. The tree-CRA was subsequently improved independently by Massey [18] and by Tsybakov and Mihailov [15]. They noticed that if after a collision the top half of the tree is empty, then a collision is certain in the bottom half. This observation was used to eliminate a number of "certain-to-contain-a collision" slots. This so-called modified-tree-CRA, combined with the simple rule for the initial transmission cited above, resulted in an improved throughput of 0.375 packets per slot.

The tree-CRA and its modification were subsequently used with a more complicated rule for the initial transmission. Suppose the unit of time is a slot, so the ith transmission slot is the interval (i,i+1]. Define a second time interval (ix, (i+1)x], where the parameter x is in units of slots. The interval (ix, (i+1)x] is called the ith arrival epoch. The new rule for the initial transmission is: transmit a new packet that was generated in the ith arrival epoch in the first slot following the CRI for new packets that were generated in the (i-1)st arrival epoch. The tree-CRA and its modification combined with this new rule for the initial transmission resulted in improved throughputs of 0.4295 and 0.4623 packets per slot, respectively [12,13,14,18]. This new rule also has the property of eliminating statistical dependency between CRI's.

Note that the original tree-CRA does not require the feedback information to distinguish between empty slots and slots containing exactly one packet; i.e., CNC binary feedback suffices for the original tree-CRA. However, the modified-tree-CRA does require the 0,1,e-ternary feedback.

1.4.3 Gallager's Algorithm and its Modification. There is an equivalent implementation of the tree-type CRAs described in Section 1.4.2 that does not actually perform the tree search but instead uses the inherent randomness of the generation times of the new packets. For example, if there is a collision in the first slot of the CRI corresponding to the ith arrival epoch, then being on the top or bottom half of the tree is equivalent to arriving in the left half or the right half of the ith arrival epoch. Furthermore, if this method is used, then the resulting CRA has the extra property that it transmits the packet in a first-come first-served (FCFS) fashion.

Gallager [16] noticed that, if this method of interval-halving is used, whenever a collision in one slot is followed by another collision in the next slot, then all information about the number of Poisson arrivals in the second half of the original collision interval has been eradicated. Accordingly, Gallager's CRA returns this second half to the as yet unexamined points of the arrival axis. Gallager incorporated this trick into the modified-tree-CRA with the new rule for the initial transmission described at the end of Section 1.4.2, and then optimized over x. He obtained a maximum throughput of 0.48711 packets per slot with x satisfying $\lambda x = 1.266$ [16,18,19].

Mosely [17] did some fine-tuning of Gallager's CRA to obtain a throughput of 0.48775. Mosely also argued that this version of Gallager's algorithm is optimum for a sub-class of FCFS algorithms.

1.4.4 Bounds on the Performance of the Optimum RAA. We defined the efficiency of a RAA for the Poisson infinite-user model as the fraction of slots devoted to exactly one packet. This is equivalent to defining the throughput η as follows. Let δ_n denote the random delay between the instant at which the nth message is generated and the instant at which its packet is successfully transmitted. Given any protocol, let η denote the supremum of all the values of λ (the global arrival rate in packets per slots) for which $\lim_{n\to\infty} E\delta_n$ exists and is finite for the said protocol, where E denotes statistical expectation. Let η_3^* denote the supremum of η over all protocols. (The subscript 3 in η_3^* refers to the 0,1,e-ternary feedback.)

Note that the throughput of any stable protocol provides a lower bound to η_3^*. Therefore, the fine-tuned version of Gallager's algorithm [17] shows that $\eta_3^* \geq 0.48775$.

On the other hand, Pippinger [21] obtained, via an information-theoretic approach, an upper bound of 0.744. His argument was improved by Hajek [24] to yield a bound of 0.7114 and subsequently improved further by Humblet [25] to show that $\eta_3^* \leq 0.704$.

Molle [22] used a "genie-aided" argument to show that $\eta_3^* \leq 0.6731$. This was improved to 0.6215 by Cruz and Hajek [23] and to 0.587 by Tsybakov and Mihailov [26]. Berger, Mehravari, and Munson [27] have conjectured that $\eta_3^* \leq 0.5254$.

Hence, at the time of this writing, η_3^* is proven to be between 0.48775 and 0.587 and is conjectured to be between 0.48775 and 0.5254.

1.5 Review of Group Testing

Group testing can be thought of as an economically efficient means of detecting defective members of large populations. Group testing was first introduced during World War II by Dorfman [32]. Dorfman's method of group testing allowed the United States Public Health Service and the Selective Service System to identify all syphilitic men called up for induction using up to 80 percent fewer blood tests than in the previously employed procedure of giving each individual a "Wasserman-type" blood test. In Dorfman's scheme after the blood samples are drawn, they are pooled in groups of n, whereupon groups rather than individuals are subjected to the test. If none of the n individuals in a group is syphilitic then the test will be negative. If, however, one or more of the individuals in the group carry the syphilitic antigen, the test will be positive. In that case, the individuals in that group must then be tested individually. Dorfman showed that on the average this scheme requires fewer blood tests and computed the most efficient size, n, for the groups.

A binomial group testing problem can be mathematically defined as follows:

DEFINITION 1.5.1 - A binomial group testing problem is concerned with classifying each of M objects as either defective or non-defective. We associate independent Bernoulli (p, q=1-p) random variables with each of the M objects, where p is the probability that an object is defective. A group test is a simultaneous test on n objects with only two possible outcomes; the choice of n is at disposal of the tester. A "good" reading indicates that all n objects are non-defective, and a "bad" reading indicates that at least one of the n objects is defective. The task is to design an efficient algorithm for correctly identifying all defective members of a population of size M.

For finite M, efficiency is defined in the sense of minimizing the expected number of group tests needed to classify all M objects. Throughout this section we make use of the following terminology:

(i) a defective set is a set known to contain one or more defective objects among its members

(ii) a non-defective set is a set known to contain no defective members

(iii) a binomial set is a set whose members each have probability p of being defective independently of one another.

Sterret [33] improved upon Dorfman's procedure by proposing that the individual members of a defective set should be tested individually only until a defective unit is found. Then the remaining units from that defective set are pooled and tested. This is continued until that particular defective set is completely analyzed.

Sobel and Groll [34] further generalized this idea by testing small subsets of a defective set rather than immediately testing individual units. The procedure proposed by Sobel and Groll has the property that at every step the remaining units (i.e., the unclassified units), are separated into two disjoint sets; a defective set and a binomial set. If at some step the defective set is empty, then a subset of the binomial set is tested. Otherwise, a subset of the defective set is tested. That is, the procedure never tests a group made up of a union of a subset of the binomial set and a subset of the defective set. This is referred to as the non-mixing rule. Furthermore, in contrast with Dorfman's and Sterrett's procedures which employed a common group test size throughout, the protocol proposed by Sobel and Groll chooses the optimum group test size at each stage. This procedure is shown to be the optimum strategy for the binomial group testing problem in the sense of minimizing the expected number of group tests needed assuming adherence to the non-mixing rule.

Table 1.5.1 displays the expected number of group test needed for the three procedures described above for selected values of population size, M, and probability of being defective, p.

Procedure / p	Dorfman	Sterrett	Sobel and Groll
0.0	1.0 1.0 1.0	1.0 1.0 1.0	1.0 1.0 1.0
0.01	1.148 1.609 2.354	1.119 1.412 1.852	1.110 1.308 1.542
0.02	1.292 2.176 3.334	1.236 1.807 2.973	1.218 1.612 2.073
0.05	1.699 3.398 5.097	1.576 3.151 4.333	1.538 2.499 3.594
0.1	2.303 4.605 6.908	2.105 4.210 6.315	2.051 3.904 5.790
0.25	3.375 6.750 10.125	3.375 6.750 10.125	3.332 6.619 9.905
0.5	4.0 8.0 12.0	4.0 8.0 12.0	4.0 8,0 12.0
1.0	4.0 8.0 12.0	4.0 8.0 12.0	4.0 8.0 12.0

This table is taken from [34]

TABLE 1.5.1 - Comparison of the expected number of group tests needed for different procedures for M=4, 8, and 12, and for selected values of p.

LECTURE 2
CONFUSION RESOLUTION ALGORITHM FOR INFINITE USER MODEL

2.1 Overview

We introduce and analyze random accessing algorithms (RAA) for a random access channel with binary feedback serving a Poisson infinite user model. After the channel and user models are presented, we describe and analyze a confusion resolution algorithm (CRA) for something/nothing (SN) feedback. Success/Failure (SF) feedback for the Poisson infinite-user model is briefly discussed, as is upper bounding the maximum achievable throughput of a random access channel with binary feedback.

2.2 User and Channel Models

Consider a large, effectively infinite, population of users attempting to exchange messages over a random multiple-access, collision-type, time-slotted, packet-switched communication channel with feedback; such a channel was fully described in Section 1.2. The generation times of information packets by the users is modeled as a Poisson point process on the time interval $T=(a,b]$ with intensity $\lambda > 0$. Therefore we can associate with each packet the time it was generated.

The goal is to design and analyze efficient CRAs and RAAs satisfying Definitions 1.2.1 and 1.2.2 for accessing the common channel. We are interested only in protocols that can be implemented in a distributed fashion by each user. On the basis of the feedback history, at the end of the kth slot the CRA designates a subset $\psi_{k+1} \subset T$. Each packet generated in ψ_{k+1} is then transmitted during the (k+1)st slot.

2.3 Something/Nothing (SN) Feedback

This type of binary feedback for the Poisson infinite user model is characteristic of public-key secure computer communications. Consider a random-access broadcast system where information packets are encoded by individual users using the public encryption key of the intended recipient. If, for example, user i is the only user who has transmitted a packet during the kth slot, then simply by listening to the channel he will be aware of the fact that the packet has been successfully transmitted. Meanwhile, any user $j \neq i$ other than the intended recipient cannot distinguish between this situation and one in which two or more messages have collided. However, for the protocol to be executable in a distributed fashion, it is necessary to inform all of the users whenever a packet has been successfully transmitted. This is done by sacrificing one slot and re-enabling (see state R below) some portions of the arrival process which have already been examined by the protocol.

Therefore, assume that at the end of kth slot each user learns the value of the feedback $Z_{SN}(k)$, where

$$Z_{SN}(k) = \begin{cases} S & \text{if } \psi_k \text{ contained} \geq 1 \text{ arrivals (something)} \\ N & \text{if } \psi_k \text{ conatined no arrivals (nothing)} \end{cases}$$

We propose a FCFS CRA deploying SN feedback which is described with the aid of the state diagram of Figure 2.3.1. The description also involves the concept of the neighbor of an interval; any interval adjacent to interval I on its right is called a neighbor of I. The states in Figure 2.3.1 dictate the following actions:

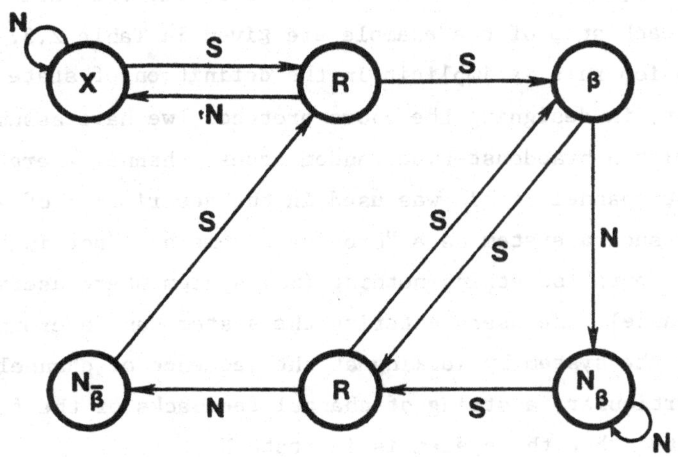

FIGURE 2.3.1 - State diagram for the CRA deploying SN feedback (infinite user model).

State x: ψ_{k+1}:= neighbor of ψ_k of length x

State β: ψ_{k+1}:= the leftmost fraction β of ψ_k

State N_β: ψ_{k+1}:= neighbor of ψ_k of length $(1-\beta)\ell_k$

State $N_{\bar{\beta}}$: ψ_{k+1}:= neighbor of ψ_k of length $\frac{1-\beta}{\beta}\ell_k$

State R: ψ_{k+1}:= ψ_k, with the exception that, if there was only one packet in the kth slot, the user who transmitted it and hence is aware that there was no collision, does not retransmit the packet he just successfully transmitted.

The algorithm is initiated in state x by letting $\psi_1 = (0,x]$.

In order to show the operation of this protocol, we apply it to the example shown in Figure 2.3.2. The protocol is initiated in state x by

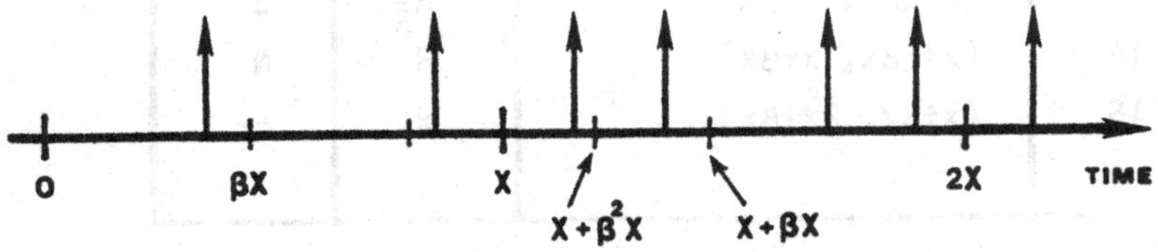

FIGURE 2.3.2 - An example of the arrival process.

designating ψ_1 = (0,x]. The feedback for each slot and the action of the protocol at each step of the example are given in Table 2.3.1. The initial transmission rule is implicit in the definition of state x. Also note that, in designing the above protocol, we have assumed that we are dealing with a braodcast-type random access channel where the users can listen to the channel. (This was used in the description of state R.) We will refer to such a system as a "broadcast something/nothing" (B-SN) in contrast to a "pure" something/nothing (SN) system where users cannot listen to the channel. The users entering the system can determine the present state of the system by looking at the sequence of channel feedbacks. In particular, a string of channel feedbacks of the form (S,S,N,N) indicates that the system is in state N_β.

i	Ψ_i	State	Z(i)
1	(0,x]	x	S
2	(0,x]	R	S
3	(0,βx]	β	S
4	(0,βx]	R	N
5	(βx, x]	$N_{\bar\beta}$	S
6	(βx, x]	R	N
7	(x, 2x]	x	S
8	(x, 2x]	R	S
9	(x, x+βx]	β	S
10	(x, x+βx]	R	S
11	(x, x+ββx]	β	S
12	(x, x+ββx]	R	N
13	(x+ββx, x+βx]	$N_{\bar\beta}$	S
14	(x+ββx, x+βx]	R	N
15	(x+βx, 2x+βx]	x	--

TABLE 2.3.1 - Ψ_i and $Z_{SN}(i)$ for the example of Fig. 2.3.2.

The following is a simple analysis of the CRA described by the state diagram of Figure 2.3.1. Let

$$\Pi_n \triangleq \frac{e^{-\lambda x}(\lambda x)^n}{n!}$$

$$P_{n,i} \triangleq \binom{n}{i}\beta^i(1-\beta)^{n-i}$$

$W_n \triangleq$ average number of states visited (i.e., number of slots employed) prior to the first return to state x, given the protocol starts in state x with a n-conflicted interval; state x does not count toward W_n either at the beginning or the end.

$B_n \triangleq$ average fraction of a basic interval which is resolved during a CRI, given the protocol starts in state x with a n-conflicted interval.

$W \triangleq$ average number of states visited (i.e., number of slots employed) between successive returns to state x.

$B \triangleq$ average number of successful transmissions between successive returns to state x.

= average number of arrivals in the resolved portion of a conflicted interval.

We have $W_0 = 0$, $B_0 = 1$, and $B_1 = 1$. Note that it takes two slots to process an interval containing one arrival. The first one is used to successfully transmit the single packet; the feedback from the second slot informs the rest of the users that the interval contained exactly one arrival. Hence, $W_1 = 1$. The recursive formulas for W_n and B_n, where $n \geq 2$, are:

$$W_n = P_{n,0}(1 + W_n) + P_{n,1}(4 + W_{n-1}) + \sum_{i=2}^{n} P_{n,i}(2 + W_i) \qquad (2.3.1)$$

$$B_n = P_{n,0}[\beta+(1-\beta)B_n] + P_{n,1}[\beta+(1-\beta)B_{n-1}] + \sum_{i=2}^{n} P_{n,i}\beta B_i \qquad (2.3.2)$$

Rewriting equations 2.3.1 and 2.3.2 and solving for W_n and B_n, we get:

$$W_n = [1+P_{n,1}(3+W_{n-1}) + P_{n,n} + \sum_{i=2}^{n-1} P_{n,i}(1+W_i)]/(1-P_{n,0}-P_{n,n}) \qquad (2.3.3)$$

$$B_n = [\beta P_{n,0}+P_{n,1}[\beta + (1-\beta)B_{n-1}] + \sum_{i=2}^{n-1} P_{n,i}\beta B_i]/(1-\beta P_{n,n} - (1-\beta)P_{n,0}) \qquad (2.3.4)$$

Averaging over n, we get:

$$W = 1 + \sum_{n=0}^{\infty} W_n \Pi_n \qquad (2.3.4)$$

$$B = \sum_{n=0}^{\infty} x\, B_n \Pi_n \qquad (2.3.6)$$

where W_n and B_n are given by equations 2.3.3 and 2.3.4. It follows that the efficiency of this CRA, as a function of β and x, is given by η_{SN} = B/W where B and W are given by equations 2.3.5 and 2.3.6. Optimizing over β and x shows that the maximum achievable throughput for the CRA is 0.2794..., corresponding to β = 0.400..., and x = 0.944... .

We believe that it is not possible to design CRAs which satisfy Definition 1.2.2 for a system with pure something/nothing feedback. However, Hajek and Van Loon [11] have recently shown that ALOHA-type retransmission policies exist which can be implemented on a random access channel with SN feedback to achieve a stable throughput of 1/e = 0.3678. However, ALOHA-type policies do not satisfy definition 1.2.2.

We also believe that it is not possible to design CRAs which satisfy definition 1.2.2 for success/failure feedback in an infinite user environment. However, note that a user who transmits in a given slot and receives a "failure" feedback for that slot knows that there has been a collision in that slot as opposed to an empty slot. This allows the implementation of an ALOHA-type scheme for SF feedback. However, ALOHA-type schemes require some kind of dynamic control policy to achieve a stable throughput. The policies introduced by Hajek and Van Loon [11] do not cover success/failure feedback.

2.4 Bounds on the Performance of Optimum RAA

The performance achieved by the protocol introduced in the last section provided lower bounds to the maximum achievable throughput of the random access channel for the corresponding binary feedback and the Poisson infinite-user model. It is also of interest to provide upper bounds to the performance of the optimum strategy for each type of feedback.

Let η^*_{SN} be the maximum achievable throughput for pure something/nothing feedback and let η^*_2 be the maximum achievable throughput for any random access channel with binary feedback.

One idea for upper bounding the efficiency of CRAs with infinite user model, suggested first by Molle [22] and then by Cruz and Hajek [23] for 0,1,e feedback, is to consider a beneficial "genie" who provides additional information to the CRA which is not provided under the original channel model. Then, it might be possible to describe optimal protocols for genie-aided systems. Since no strategy that requires the genie's information to operate is implementable without the genie, the performance of optimal genie-aided protocols will upper bound the performance of any unaided protocol. We introduce similar arguments to provide upper bounds to η^*_2 and η^*_{SN}.

THEOREM 2.4.1 - $\eta^*_2 \leq 0.587$.

PROOF - Consider a beneficial genie who turns the binary feedback of a random access channel into a 0,1,e feedback and provides the improved feedback to all of the users. Clearly the optimum protocol for the 0,1,e feedback will perform at least as well as the optimum protocol for any binary feedback. Moreover, the performance of the optimum protocol for the 0,1,e feedback is bounded above by 0.587 [26].

Hence, $\eta^*_2 \leq \eta^*_3 \leq 0.587$. QED

CONJECTURE 2.4.1 - $\eta^*_2 \overset{?}{\leq} 0.5254$

ARGUMENT - The same reasoning given for the above theorem combined with the conjectured upper bound of 0.5254 [27] for η^*_3 shows that $\eta^*_2 \leq \eta^*_3 \overset{?}{\leq} 0.5254$.

THEOREM 2.4.2 - $\eta^*_{SN} \leq 0.50$.

PROOF - Suppose that, in the event that the feedback random variable has taken the value S for the interval $(0,x]$, a beneficial genie provides for us the exact location t_1 of the first arrival in $(0,x]$. There may or may not be packets beyond t_1; the genie does not provide us with any information about this matter. Hence, after the genie's information has been provided, the arrival points in $(t_1,x]$ are again Poisson distributed with parameter λ. Therefore, after transmitting the packet generated at t_1 in the next slot, we investigate a new basic interval of length x, $(t_1, t_1+x]$, without paying attention to the feedback generated from the slot containing the packet generated at t_1. Doing so yields an empty interval with probability e^{-x}, and another interval with one or more arrivals with probability $1-e^{-x}$. In the event of another S feedback, the genie will help us again. If we get a N feedback, we go to the next basic interval of length x. It is "clear" (see [22]) that the above procedure makes optimum use of the information provided by the genie. Therefore, the fraction of slots that yield successfully transmitted packets using this procedure serves as an upper bound to the optimum efficiency η^*_{SN} attainable without the genie's help. The fraction is easily seen to be:

$$\frac{1-e^{-x}}{e^{-x}+2(1-e^{-x})} = \frac{1-e^{-x}}{2-e^{-x}}$$

This is maximized at $x = \infty$, and the maximum value is 0.50. Hence, $\eta^*_{SN} \leq 0.50$. QED

LECTURE 3

CONFUSION RESOLUTION ALGORITHMS FOR FINITE USER MODEL AND GENERALIZED GROUP TESTING

We introduce and analyze CRAs for random access channels with binary feedback and finitely many users. We introduce the idea of the generalized binary binomial group testing problem and use it to design efficient confusion resolution CRAs. For both SN and SF feedback we study two different classes of algorithms - one based on binary tree search techniques, and the other based on generalized group testing methods. The binary tree search techniques are based on the original ideas of Hayes [36] and Capetanakis [12,13,14]. It turns out that the protocols based on group testing methods are more efficient than the ones based on binary tree search techniques. It is shown that group-testing-based random accessing algorithms are superior to TDMA for small values of the probability p that a user has a message to transmit, and perform at least as well as TDMA for higher values of p. In particular, it is shown that the proposed algorithm for SN feedback is more efficient than TDMA for $p < p^*_{SN} = (3-\sqrt{5})/2$ and adapts itself to TDMA for $p \geq p^*_{SN}$. We also discuss bounds on the performance of the optimum strategies.

3.1 Channel and User Models

Consider a finite population of M users attempting to exchange messages over a random multiple-access, collision-type time-slotted, packet-switched communication channel with feedback.

Users generate packets independently of one another. Each user can have at most one packet in the process of being transmitted or retransmitted and at most one waiting to be processed. Packets arriving during the present confusion resolution interval (CRI) must wait until the next CRI to be transmitted. The probability that a user who does not yet have a packet waiting will generate one during the next slot is a constant ρ. (We assume ρ to be a known system parameter. However, if ρ is unknown, its value can be estimated by observing the transmission process during an appropriate length of time.) Hence, p, the probability that user will generate a packet during a CRI of length t is given by

$$p = 1 - (1-\rho)^t. \tag{3.1.1}$$

Assume the arrival axis to be divided into so called "arrival segments"; the length of kth segment is defined to be equal to that of the (k-1)st CRI.

All of the protocols described in this chapter operate on the arrival process segment by segment; that is, they process all of the arrivals in a given segment before attending to the next segment. All of the packets generated in the kth segment will be successfully transmitted in the kth CRI. The operation and performance of all of the protocols we shall introduce will depend on the statistics of the arrival process only through p. Hence, other probabilistic models can be used to model a finite user system as long as the value of p is computable from the model [12,14,36-38].

Throughout this lecture we will compare the performance of the protocols against each other and against TDMA. Our measure of performance is given by

$W(M;p) \triangleq W(M)$ = Average number of slots needed to resolve an arrival segment, for a system with M users and given value of p.

= Average length of a CRI, for a system with M users and a given value of P.

Our task is to design CRAs and RAAs to minimize W(M). Observe that achieving a smaller value of W(M) for fixed p is tantamount to satisfactorily handling a larger underlying packet generation rate. Note that W(M) = M for TDMA for all values of p, $0 \leq p \leq 1$.

We will use the simple initial transmission rule of transmitting a new packet in the first slot following the CRI that was in progress when the packet was generated.

3.2 Generalized Binary Binomial Group Testing

The concept of group testing and a number of efficient group testing procedures were reviewed in Section 1.5. We now introduce the idea of a generalized group testing problem which will be used later for developing efficient CRAs for a random access channel with a finite number of users and a binary feedback.

DEFINITION 3.2.1 - A generalized binary binomial group testing problem is concerned with classifying each of M objects as either defective or non-defective. We associate independent Bernoulli (p, q=1-p) random variables with each of the M objects, where p is the probability that an object is defective. Let R(n) be the random variable representing the number of defective objects in a group of size n. A generalized group test T(a,b;n) is a simultaneous test of n objects with only two mutually exclusive possible outcomes; the choice of n is at the disposal of the tester. A "good" reading indicates that $a \leq R(n) \leq b$ and a "bad" reading indicates otherwise, where integers a and b satisfy $0 \leq a \leq b \leq n$. The task is to design efficient algorithms for correctly classifying all members of a population of size M.

Sometimes it will be more convenient to use the notation T(a,b;G) to represent a generalized group test performed on group G. In that case the random variable R(G) represents the number of defective objects in G. For finite M, efficiency is defined in the sense of minimizing the expected number of group tests needed to classify all m objects. The definitions of the defective set, non-defective set, and the binomial set are the same as those used in Section 1.5. Note that, if a = 0 and b = 0, then the definition of a generalized group testing problem coincides with that of the conventional group testing problem given in Section 1.5.

The idea of using group testing methods for developing efficient techniques for accessing a multiple-user system was considered by Berger [39], Pippenger [40], and Towsley and Wolf [31]. In a recent paper, Towsley and Wolf [31] applied conventional group testing methods to design efficient reservation protocols for multi-access communication systems. Our goal is to use generalized group testing techniques to design efficient CRAs for random multiple access systems deploying binary feedback.

By judicious specification of the integers a and b, a generalized group test can be put into one-to-one correspondence with any one of the three types of binary feedback described in Section 1.3. The appropriate relations are:

$a = 0,\ b = 0;\quad T(0,0;n) \leftrightarrow$ Something/Nothing Feedback

$a = 0,\ b = 1;\quad T(0,1;n) \leftrightarrow$ Conflict/No Conflict Feedback

$a = 1,\ b = 1;\quad T(1,1;n) \leftrightarrow$ Success/Failure Feedback.

3.3 Something/Nothing (SN) Feedback

We now propose and analyze three different random accessing algorithms for a random access channel with finite number, M, of users deploying a SN feedback. Two of them are based on binary tree search procedures and are discussed in Sections 3.3.1 and 3.3.2. Tree search protocols were first used for MA systems by Hayes [36] and Capetanakis [12,13,14]. These procedures are easy to implement and their implementation is independent of the probability, p, that a user will have a message to send. However, they require the number of users, M, to be a power of 2. Section 3.3.3 contains the description of the third procedure which is based on generalized group testing. It outperforms the protocols based on tree search methods and does not require the number of users to be a power of 2. However, its implementation depends on p. We will refer to these three procedures as A1, A2 and A3, respectively.

3.3.1 Procedure A1: Binary Tree Search. Consider a random multiple-access channel with $M = 2^K$ users, where the channel and the users satisfy the model given in Section 3.1. The users are numbered $1,2,\ldots,2^K$ arbitrarily before the start of a CRI. The indexing does not change throughout a CRI. It can, however, be changed between any consecutive pair of CRIs. Procedure A1, which is described below, is not a first-come first-served CRA. The packets are successfully transmitted according to the index of the users generating the packets. Consider the users as 2^K leaves of a binary tree. Such a tree for K=4 is shown in Figure 3.3.1.1. The level of a node is defined as the number of branches between the node and the root node. The root node is at the level zero. Nodes of the tree are labeled n_{ij}, $1 \leq j \leq 2^i$, $0 \leq i \leq K$, where j corresponds to a particular node at level i, and there are 2^i nodes of level i. CRA A1 can be described as:

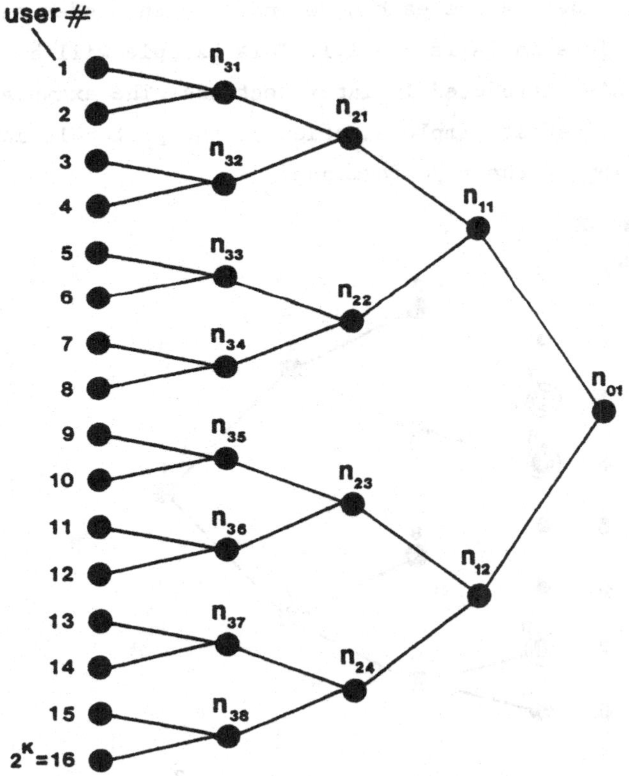

FIGURE 3.3.1.1 - Binary Tree for K=4.

PROCEDURE A1: After an S feedback (if any), all transmitters on the top half of the tree retransmit in the very next slot; those on the bottom half retransmit in the next slot after the confusion (if any) among those on the top half has been resolved. Continue in this manner until either there is no more confusion or you have reached the Kth level of the tree, at which point individual users will be transmitting their packets.

This protocol can be implemented by each of the users simply by growing the tree from the root node. Note that a confusion among users may be resolved without visiting all the nodes of the tree. This fact and other properties of this CRA can be easily shown by an example: Let $M = 2^3 = 8$. Suppose the initial confusion was among transmitters #3, #4, and #7. The corresponding binary tree is shown in Figure 3.3.1.2 and the feedback for each slot and the action of users at each step is given in Table 3.3.1.1. In Figure 3.3.1.2, numbers above each node indicate the index of corresponding time slots in Table 3.3.1.1. This example will be applied to describe other CRAs introduced in later sections. The example was designed solely to illustrate implementation of the protocols and not to provide any indication of their performance.

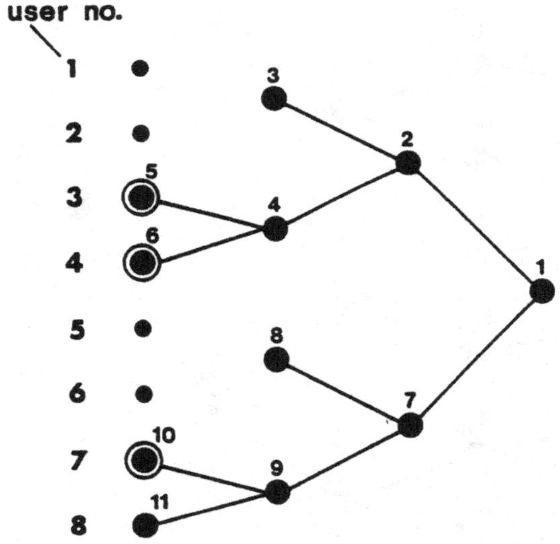

FIGURE 3.3.1.2 - Example for procedure A1, $M=2^3=8$, users number 3, 4 and 7 have packets.

Slot Number	Users Transmiting	Feedback
1	3,4,7	S
2	3,4	S
3	---	N
4	3,4	S
5	3	S
6	4	S
7	7	S
8	---	N
9	7	S
10	7	S
11	---	N

TABLE 3.3.1.1 - Channel Feedback and the Action of Protocol A1 for the Example of Figure 3.3.1.1.

Note that an "S" feedback resulting from a "transmission of level K" implies a successful transmission, where a "transmission of level i" means a transmission corresponding to a node on the ith level of the tree. In order to study some of the properties of algorithm A1 and in order to introduce a scheme for implementation of A1, we introduce the following categories of slots:

(type I slots) = (slots which resulted in an N feedback)

(type II slots) = (slots associated with transmission of level K resulting in an S feedback)

(type III slots) = (slots associated with transmission of level < K resulting in an S feedback)

Note that,

(type I slots) U (type II slots) ↔ (terminal nodes of the tree)

and

(type III slots) ↔ (intermediate nodes of the tree).

A confusion in a slot is resolved when the number of subsequent type I and type II slots exceeds by 2 the number of subsequent type III slots.

This is because the number of terminal nodes of a complete binary tree is one higher than the intermediate nodes; a complete binary tree is one in which every intermediate node has 2 branches. This property suggests the following simple way of implementing protocol A1. This method of implementation was first introduced by Gallager and then by Massey [18] for the tree protocol introduced by Capetanakis [12,13,14].

IMPLEMENTATION OF A1: Each user has two counters:

i) Each user must know when original confusion (if any) has been resolved, as this determines when new packets may be sent. For this, the first counter C1 is set to 1 prior to the first slot, then incremented by 1 for each subsequent type III slots and decremented by 1 for each subsequent type I or type II slot; when C1 reaches zero, the original confusion is resolved.

ii) Each user should also be able to implement the protocol A1 by himself, as this is necessary for the users to determine when to retransmit. For this, counter C2 is initialized to zero and then is updated as follows:

C2:=1 following a type III slot where the user was on the bottom half of the tree,

C2:=C2 + 1(C2>0) following any other type III slot,

C2:=C2 - 1(C2>0) following a type II or I slot.

Whenever C2 is equal to zero, the user retransmits in the next slot. (1(E) is the indicator function of event E.) To illustrate this implementation, we have applied it to the example of Figure 3.3.1.2 and the results are shown in Table 3.3.1.2. It remains to analyze the performance of algorithm A1. Define

$W(M;p) = W(M) \overset{\Delta}{=}$ average number of slots needed to resolve an arrival segment, for a system with $M=2^K$ users and a given value of p.

= average length of CRI, for a system with $M=2^K$ users and a given value of p. (3.3.1.1)

and let

$$y_{ij} \overset{\Delta}{=} \begin{cases} 1 & \text{if node } n_{ij} \text{ is visited} \\ 0 & \text{if otherwise} \end{cases} \qquad (3.3.1.2)$$

Slot Number	Users Transmiting	Feedback	Slot Type	First Counter	Second Counter #3	#4	#7
				1	0	0	0
1	3,4,7	S	III	2	0	0	1
2	3,4	S	III	3	1	1	2
3	---	N	I	2	0	0	1
4	3,4	S	III	3	0	1	2
5	3	S	II	2	*	0	1
6	4	S	II	1		*	0
7	7	S	III	2			1
8	---	N	I	1			0
9	7	S	III	2			0
10	7	S	II	1			*
11	---	N	I	0			

* means: successful transmission

TABLE 3.3.1.2 - Implementation of protocol A1 for the example of Fig. 3.3.1.2.

Then

$$W(M) = E \sum_{i=0}^{K} \sum_{j=1}^{2^i} y_{ij} = \sum_{i=0}^{K} \sum_{j=1}^{2^i} P(i,j) \qquad (3.3.1.3)$$

where $P(i,j)$ is the probability of visiting node n_{ij}. But any node at level i is visited if and only if there are one or more packets generated by the users on the leaves of a sub-tree whose root node is at level (i-1). Hence,

$$P(i,j) = \begin{cases} 1 - (1-p)^{2^{K-i+1}} & ; \quad i > 0 \\ 1 & ; \quad i = 0 \end{cases} \qquad (3.3.1.4)$$

Using this expression for $P(i,j)$ in equation (3.3.1.3), we get

$$W(M) = 1 + \sum_{i=1}^{K} \sum_{j=1}^{2^i} \left[1 - (1-p)^{2^{K-i+1}}\right]$$

$$= 1 + \sum_{i=1}^{K} \left[2^{i}\ 1 - (1-p)^{2^{K-i+1}} \right] \qquad (3.3.1.5)$$

The values of W(M) for $M=2^6=64$ users have been illustrated in Figure 3.3.1.3 as a function of p. For comparison purposes, the corresponding performance of TDMA, i.e. W(M) = M, is also illustrated in Figure 3.3.1.3. Note that protocol A1 is superior to TDMA for small values of p; however, it performs much worse than TDMA for higher values of p.

3.3.2 Procedure A2: Improved Binary Tree Search Method. The performance of algorithm A1 can be improved by using a trick similar to that first suggested by Massey [18] to improve the performance of the Capetanakis tree-CRA. This trick was briefly discussed in Section 1.4.2.

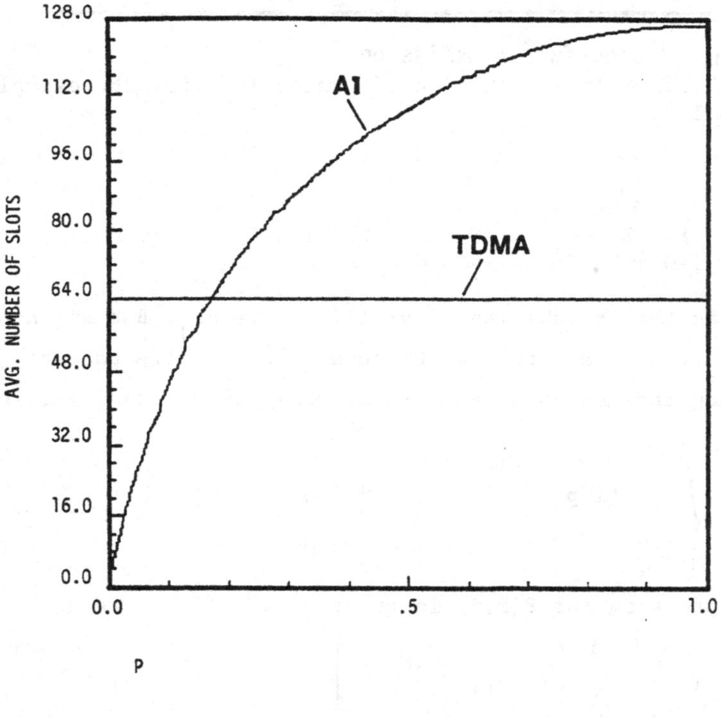

FIGURE 3.3.1.3 - Performance of procedure A1 and TDMA for M=64.

Algorithm A2 is the improved version of A1 and is described below.

PROCEDURE A2: Same as A1 except that, when the feedback indicates that a slot in which a set of transmitters on the top half of a subtree have transmitted is in fact empty, then the transmitters on the bottom half of the subtree are further divided into two parts, say the top of the bottom and the bottom of the bottom. In the next slot, the top of the bottom retransmits. Users in the bottom half of the bottom retransmit in the next slot after the confusion (if any) among those in the top half of the bottom is resolved (subject to the above exception).

In order to illustrate the operation of this protocol, we apply it to the same example that was used for protocol A1. The example is repeated in Figure 3.3.2.1 and the operation of the protocol is summarized in Table 3.3.2.1. Note that procedure A2 performs better in the example by skipping two of the nodes which A1 would have visited.

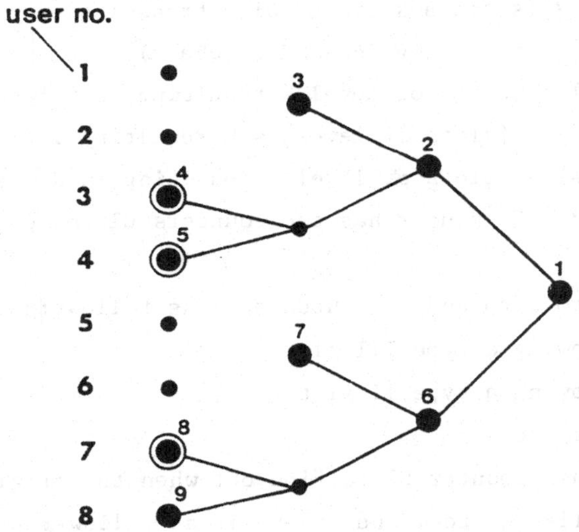

FIGURE 3.3.2.1 - Example for procedure A2; $M=2^3=8$, users number 3, 4, and 7 have messages.

Slot Number	Users Transmiting	Feedback
1	3,4,7	S
2	3,4	S
3	---	N
4	3	S
5	4	S
6	7	S
7	---	N
8	7	S
9	---	N

TABLE 3.3.2.1 - Channel feedback and the action of Protocol A2 for the example of Fig. 3.3.2.1.

In order to generate a scheme for implementation of protocol A2 similar to that for protocol A1, we introduce the following categorization of slots:

(type I slots) = (slots associated with transmission of levels < K resulting in an N feedback)

(type II slots) = (slots of level K resulting in S feedback)

(type III slots) = (slots of levels < K resulting in S feedback)

(type IV slots) = (slots of level K resulting in N feedback)

IMPLEMENTATION OF A2: Each user has two counters C1 and C2, and one flag F1.

i) Initialize F1 to zero and then update it as follows:

F1:=1 following a type III slot

F1:=0 following a type II slot

F1:=F1 otherwise

ii) Stations will use counter C1 to find out when the original collision is resolved. Initialize C1 to 1 and update it as follows:

C1:=C1+1 following type III slots

C1:=C1 following type I slots when F1 is set to 1

C1:=C1-1 following type II slots, or type IV slots, or type I slots when F1 is set to zero.

When C1 reaches zero, the original collision has been resolved.

iii) Each station can implement protocol A2 by using counter C2. Initialize C2 to zero and update it as follows:

C2:=1 Following type III slots where the user was on the bottom half of the tree,

C2:=C2+1(C2>0) following type III slots,

C2:=C2-1(C2>0) following type II, or type I when F1=0, or type IV, of type I when F1=1 and C2=1 and the user is on the top half of the tree.

C2:=C2 otherwise

When C2 reaches zero, the user retransmits in the following slot. To illustrate this implementation, we have applied it to the example of Figure 3.3.2.1 and the results are shown in Table 3.3.2.2.

It remains to analyze the performance of protocol A2. Define W(M),

Slot Number	Users Transmiting	Feedback	Slot Type	F1	C1	C2		
						#3	#4	#7
				0	1	0	0	0
1	3,4,7	S	III	1	2	0	0	1
2	3,4	S	III	1	3	0	0	1
3	-	N	I	1	3	0	1	2
4	3	S	II	0	2	*	0	1
5	4	S	II	0	1		*	0
6	7	S	III	1	2			1
7	-	N	I	1	2			0
8	7	S	II	0	1			*
9	-	N	IV	0	0			

* Indicates successful transmission

TABLE 3.3.2.2 - Implementation of protocol A2 for the example of Fig. 3.3.2.1.

Y_{ij}, and P(i,j) as given in the previous section. We are interested in W(M) which is given by

$$W(M) = \sum_{i=0}^{K} \sum_{j=1}^{2^i} P(i,j) \tag{3.3.2.1}$$

where P(i,j) is the probability of visiting node n_{ij}. Therefore, the main task is to calculate P(i,j). Recall that the index i corresponds to a particular level of the tree. Each node at level i corresponds to possible transmissions by 2^{K-i} users. We will refer to a user with a message to transmit as an "active user." To evaluate P(i,j), we introduce the symbols f_i, r_i, q_i, and t_i which are defined as:

$f_i \triangleq P$ (there is at least one active user within 2^{K-i} users),

$r_i \triangleq P$ (there is at least one active user within one half of a set S of 2^{K-i+1} users, given there is at least one active user within S),

$q_i \triangleq 1 - r_i$

$t_i \triangleq q_i + r_i f_i$

These quantities can easily be expressed in terms of p,

$$f_i = 1 - (1-p)^{2^{K-i}} \quad \text{for } i=0,1,2,\dots,K-1 \tag{3.3.2.2}$$

$$r_i = f_i/f_{i-1} \quad \text{for } i=1,2,3,\dots,K-1 \tag{3.3.2.3}$$

$$q_i = 1 - f_i/f_{i-1} \quad \text{for } i=1,2,3,\dots,K-1 \tag{3.3.2.4}$$

$$t_i = 1 - \frac{f_i}{f_{i-1}}(1-f_i) \quad \text{for } i=1,2,3,\dots,K-1 \tag{3.3.2.5}$$

The values of P(i,j) for $i \leq j \leq 2^i$ for $0 \leq i \leq 3$ are shown in terms of f_i, r_i, q_i, and t_i in Figure 3.3.2.2. Each of the intervals in Figure 3.3.2.2 is labeled with the node that it corresponds to, and the arrows indicate the possible paths for reaching each node. The general pattern for expressing P(i,j) in terms of f_i, r_i, q_i, and t_i is evident from the Figure 3.3.2.2. We deduce that

$$\begin{aligned} W(M) = 1 &+ f_0(1 + r_1) \\ &+ f_0(r_1 + r_1 r_2 + t_1 + t_1 r_2) \\ &+ f_0(r_1 r_2 + r_1 r_2 r_3 + r_1 t_2 + r_1 t_2 r_3 + \end{aligned}$$

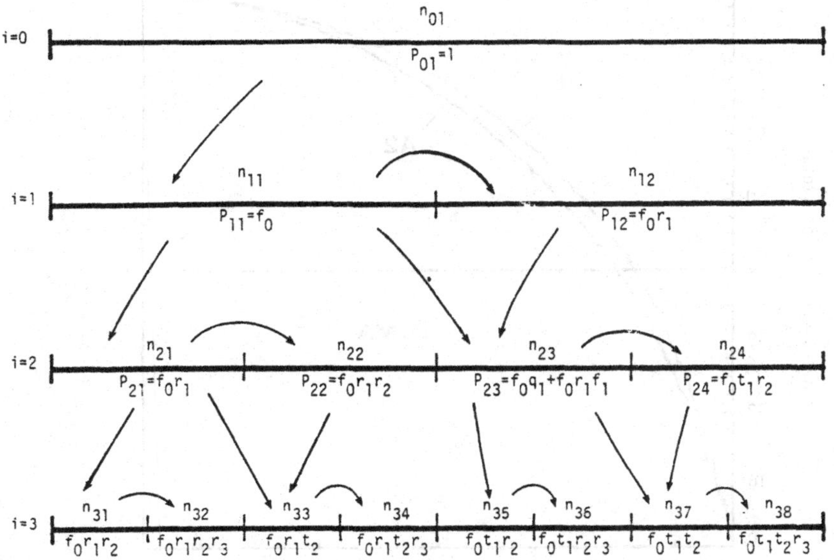

FIGURE 3.3.2.2 - $P(i,j)$ for $1 \leq j \leq 2^i$ for $p \leq i \leq 3$ for Procedure A2.

$$t_1r_2 + t_1r_2r_3 + t_1t_2 + t_1t_2t_3)$$

$$+ \ldots$$

$$= 1 + f_0(1 + r_1)$$

$$+ f_0(1 + r_2)(r_1 + t_1)$$

$$+ f_0(1 + r_3)(r_2 + t_2)(r_1 + t_1)$$

$$+ \ldots$$

It follows that

$$W(M) = 1 + f_0 \sum_{i=1}^{K} (1 + r_i) \prod_{j=1}^{i-1} (r_j + t_j)$$

where f_i, r_i, and t_i are given by Equations (3.3.2.2)-(3.3.2.5). Note that $r_K = 1$ because given a confusion at a node from level (K-1), the corresponding pair of nodes at level K are visited with probability 1.

The values of W(M) for procedure A2 for $M=2^6=64$ users along with the corresponding values of procedure A1 and TDMA are illustrated in Figure 3.3.2.3 as a function of p. The results in Figure 3.3.2.3 show that protocol A2 outperforms A1 for all values of p, in return for some

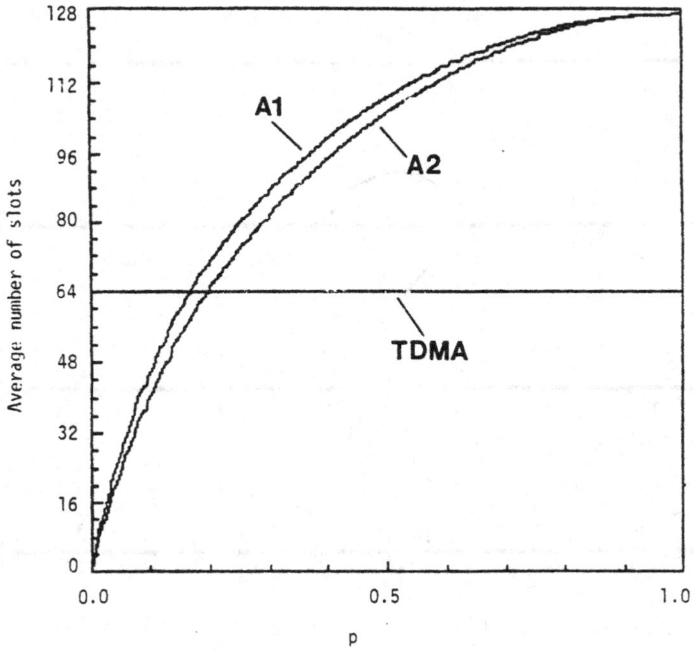

FIGURE 3.3.2.3 - Performance of procedures A1, A2 and TDMA for M=64.

complexity in the implementation. However, for large values of p it still performs much worse than does TDMA.

3.3.3 Procedure A3: Generalized Group Testing Method. The two CRAs based on binary tree search methods studied in the last two Sections had the advantage of possessing simple implementations independent of value of p. However, they required the number of users, M, to be a power of 2. A more serious indictment is that their performance was much worse than TDMA for large values of p. We now study a third CRA which we shall refer to as protocol A3. This protocol, which is based on group testing techniques, is superior to both A1 and A2 for all values of p and adapts itself to TDMA for large values of p. The other advantage of A3 over A1 and A2 is that it does not require the number of users to be a power of 2. However, its implementation depends on the value of p.

Let a=b=0 in Definition 3.2.1 of a generalized group test. The resulting group-test, T(0,0;G), corresponds to a SN feedback in our communication problem. Moreover, with a=b=0, the generalized test coincides with the conventional group-test studied by Dorfman [32], Sterrett [33], and Sobel and Groll [34]. The procedure A3 is basically the same group testing technique introduced by Sobel and Groll [34], but it has been put into a form which can be used as a CRA. We will describe it first as a group testing method and then reformulate it to fit a CRA. In what follows we will use the same definitions and terminology used by Sobel and Groll [34].

The definition of a defective set, a non-defective set, and a binomial set are as in Section 1.5. This group testing procedure has the property that, at each step, all of the unclassified units are divided into a defective set and a binomial set. Let n represent the number of unclassified units at some step of the procedure, and let m be the size of the defective set at some step. Hence, the binomial set is of size n-m.

Let random variable Y be the number of defective units in a defective set of size m and random variable Z be the number of defective units in a random sample of size x taken from a defective set of size m. Then

$$P(Y=y|Y\geq 1) = \frac{\binom{m}{y} p^y q^{m-y}}{1-q^m} \qquad (3.3.3.1)$$

and

$$\begin{aligned} P(Z=0|Y\geq 1) &= \sum_{y=1}^{m-x} p(Y=y|y\geq 1)P(A=0|Y=y) \\ &= \sum_{y=1}^{m-x} \frac{\binom{m}{y} p^y q^{m-y}}{1-q^m} \frac{\binom{m-x}{y}}{\binom{m}{y}} \\ &= \frac{q^x \sum_{y=1}^{m-x} \binom{m-x}{y} p^y q^{m-x-y}}{1-q^m} \\ &= \frac{q^x(1-q^{m-x})}{1-q^m} = \frac{q^x - q^m}{1-q^m} \end{aligned} \qquad (3.3.3.2)$$

Before describing the procedure, consider the following simple fact given in a more general setting by Sobel and Groll [34].

FACT 3.3.3.1 - Let a=0 and b be any integer in Definition 3.2.1 of a generalized group-test. Let S1 and S2 be two disjoint set of objects. If a test $T(0,b;S1 \cup S2)$ produces a "bad" reading and another test $T(0,b;S1)$ also produces a "bad" reading, then the conditional distribution of objects in S2 is the same as the original binomial distribution.

If at some step of the procedure the defective set is empty, we call that an "H-situation"; if it is not empty we call it an "F-situation". Let x(H) and x(F) be integers to be optimized later. The group testing procedure can now be described as follows: If you are faced with an H-situation, then test a group of size x(H) from the binomial set; If you are faced with an F-situation, then test a group of size x(F) from the defective set; After each group-test, the defective and binomial sets are updated according to the result of the test using Fact 3.3.3.1.

It is perhaps better to imagine the operation of this group testing method in terms of the following pair of recursive equations. Let H(n) be the average number of group-tests remaining to be performed if the defective set is empty and the binomial set is of size n, and let F(m) be the average number of group-tests needed to reach for the first time an H situation if the defective set is of size m. Then the recursive equations describing the protocol are:

$$H(n) = 1 + \min_{1 \le x \le n}\{q^x H(n-x) + (1-q^x)[F(x) + \sum_{i=1}^{x} \frac{q^{i-1}(1-q)}{1-q^x} H(n-i)]\} \qquad (3.3.3.3)$$

and

$$F(m) = 1 + \min_{1 \le x \le m-1} \{\frac{q^x - q^m}{1-q^m} F(m-x) + \frac{1-q^x}{1-q^m} F(x)\} \quad ; \quad \text{for } m \ge 2 \qquad (3.3.3.4)$$

The boundary conditions are

$$H(0) = 0 \qquad (3.3.3.5)$$

$$F(1) = 0 \qquad (3.3.3.6)$$

REMARK 1: The constant 1 in Equations (3.3.3.3) and (3.3.3.4) represents the carrying out of the very next group-test of size x, and the expression in the braces is the conditional average number of remaining group-test given x.

REMARK 2: The form of $F(m)$ in (3.3.3.3) and (3.3.3.4) is justified if we make use of Fact 3.3.3.1 with $b=0$, S1 as the defective set of size m, and S2 as a subset of size $x<m$.

REMARK 3: Let integers $x(H(n))$ and $x(F(m))$ represent the value of x which achieves the minimum in (3.3.3.3) and (3.3.3.4), respectively. These integers implicitly define the procedure. They can be found by solving (3.3.3.3) and (3.3.3.4) recursively, starting with the boundary conditions given by (3.3.3.5) and (3.3.3.6).

REMARK 4: Note that if $m>1$, the procedure "breaks down" the defective set continually until a defective object is identified.

REMARK 5: If $m>1$, it is assumed that a subset of size x with $1 \leq x \leq m-1$ will be taken from the defective set without mixing it with units from the binomial set. It follows from (3.3.3.3)-(3.3.3.6) that any lack of optimality can arise only from this "non-mixing rule."

It remains to reformulate this group testing procedure to obtain a CRA. Index the users 1,2,3,...,M. As was the case for protocols A1 and A2, this indexing can be done in any fashion. The only restriction is that the indexing cannot be changed throughout a CRI. It can, however, be changed between any consecutive pair of CRIs. The packets will be successfully transmitted in the order of the indexing of the users generating them.

A defective unit in the group testing problem corresponds to an active user in the communication problem, i.e., a user with a message to send. A classified user is one who is known either not to be active or to have transmitted his packet successfully. Moreover, a defective set corresponds to a set of users known to contain at least one active user, and a binomial set is a set of users whose members are active with probability p, independently of one another. Then protocol A3 can be described as follows:

PROCEDURE A3: If we have an H-situation, then the $x(H)$ users from the binomial set with the least indices are enabled; If we have an F-situation, then the $x(F)$ users from the defective set with the least indices are enabled. After each slot, binomial and defective sets are updated using the channel feedback.

To illustrate the operation of this protocol, we apply it to the same example that was used to illustrate protocols A1 and A2. Therefore, assume that M=8 and that users number 3, 4, and 7 have messages to send. Also assume that p = 0.15. Optimum values of x(G(n)) and x(F(m)) for p = 0.15 are shown in Table 3.3.3.1; these values are taken from [34]. The operation of protocol A3 for this example is summarized in Table 3.3.3.2.

To study the performance of protocol A3, we are interested in evaluating W(M). But we can see from the definition of H(n) that W(M) = H(M). Figure 3.3.3.1 illustrates the values of W(M) as a function of p for M=64. Values of W(M) for procedures A1, A2, and TDMA also are shown in Figure 3.3.3.1 for comparison purposes. As Figure 3.3.3.1 illustrates, procedure A3 is superior to A1 and A2 for all values of p, outperforms TDMA for small values of p, and adapts itself to TDMA for larger values of p. Let $P^*_{SN}(M)$ be the value of p at which protocol A3 switches itself to TDMA. It will be shown in the next section that $P^*_{SN}(M) = P^*_{SN} = (3-\sqrt{5})/2$ independently of number of users M.

3.3.4 On the Performance of the Optimum Strategy for SN Feedback. The performance of procedures A1, A2, and A3 provide upper bounds to the performance of the (as yet unknown) optimum strategy for the SN feedback in the sense of minimizing the average number of slots. We now look

k	x(H(k))	x(F(k))
1	1	
2	2	1
3	3	1
4	4	2
5	5	2
6	6	2
7	4	2
8	4	3

TABLE 3.3.3.1 - Values of x(H(k)) and x(F(k)) for p = 0.15.

Slot Number	Situation	x	Users Transmiting	Feedback
1	H(8)	4	3,4	S
2	F(4)	2	---	N
3	F(2)	1	3	S
4	H(5)	5	4,7	S
5	F(5)	2	4	S
6	F(2)	1	4	S
7	H(4)	4	7	S
8	F(4)	2	---	N
9	F(2)	1	7	S
10	H(1)	1	---	N

TABLE 3.3.3.2 - operation of procedure A3 for M=8 when users number 3, 4, and 7 have messages and p = 0.15.

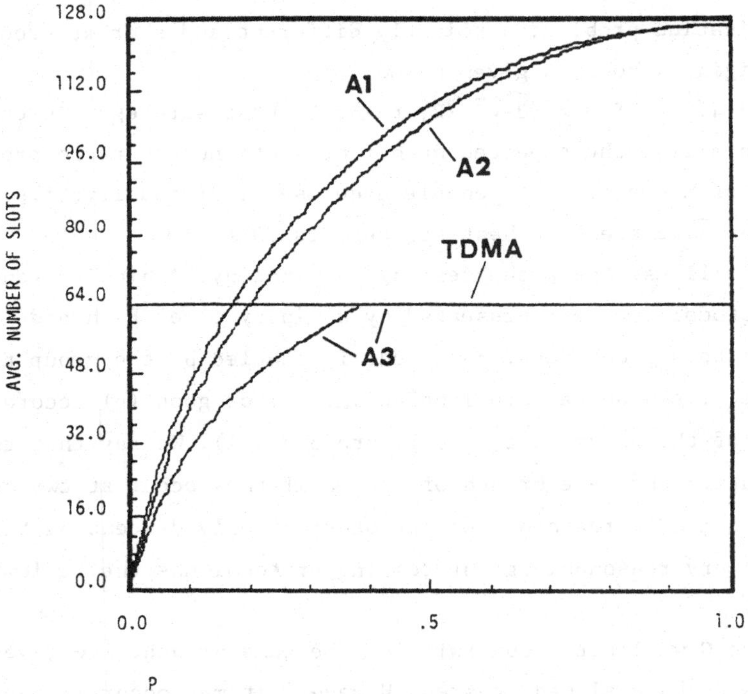

FIGURE 3.3.3.1 - Performances of protocols A1, A2, A3, and TDMA for M=64.

at some properties of procedure A3 and proceed to provide a lower bound to the performance of the optimum strategy.

Let $P^*_{SN}(M)$ be the value of p at which procedure A3 switches to TDMA; i.e., the corresponding group testing procedure tests groups of size 1 only. One of the interesting properties of procedure A3 is that $P^*_{SN}(M) = P^*_{SN} = (3-\sqrt{5})/2$ independently of the number of users.

THEOREM 3.3.4.1 - If $1 \leq m \leq n$ and $1 \geq p \geq p^*_{SN} = (3-\sqrt{5})/2$ then

$$x(H(n)) = 1$$

$$H(n) = 1$$

$$F(m) = \frac{p}{q} + \frac{1-q^{n-1}-mq^m}{1-q^m}$$

The proof is by induction on n and is given by Sobel and Groll [34].

In a paper written in 1960, Unger [35] showed that the range of p for which there exists group testing methods such that the expected number of tests is less than M (for a population of size M) is $0 \leq p < (3-\sqrt{5})/2$. This result is stated in the following theorem for the communication problem; a slightly different and shorter proof than Unger's original proof is given below.

THEOREM 3.3.4.2 - If $p \geq (3-\sqrt{5})/2$, then the best strategy, in the sense of minimizing the expected number of slots necessary to process a group of M users, is to enable each user individually; i.e., if $p \geq (3-\sqrt{5})/2$ then the best strategy is TDMA.

PROOF - We will use the group testing terminology. A binary group testing protocol can be represented by a binary tree. Each node corresponds to a group test. Each node is labeled by the group tested at the node. The branches are labeled bad (b) or good (g) according to the result of the group test (see Figure 3.3.4.1). We say that two groups occur on the same branch of a tree if they occur at two nodes, one of which can be reached from the other by only descending through the tree. Every reasonable group testing protocol has the following properties:

1. A group G will not occur twice on the same branch; i.e., we will not test groups we have already tested. However, it may occur in many places on the test tree.

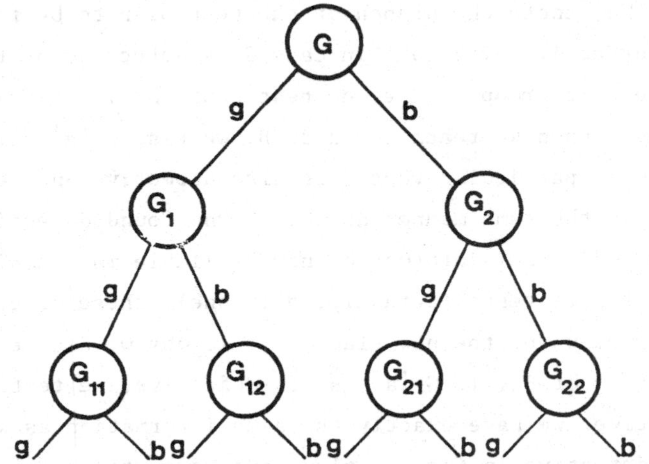

FIGURE 3.3.4.1 - Group Testing Procedure as a Binary Tree.

2. Let G be the group corresponding to node B of the test tree. No element of G occurs in any of the groups on the "good" branch of node B; i.e. nothing is gained by adding users known to be non-defective to any group.

3. Let G be the group corresponding to node B of the test tree. Then no group $G' \supset G$ occurs on either branch beneath node B. Indeed, if G is defective, then G' is known to be defective, and if otherwise, see property 2 above.

4. A test is not performed if its outcome can be inferred from previous results.

If a protocol does not satisfy properties 1-4, we can change it into one which does by removing elements from groups and omitting unnecessary tests. The number of tests needed to classify any sample is certainly not increased by these modifications.

We prove the theorem by modifying an arbitrary protocol which satisfies properties 1-4 and contains a group-test of more than 1 element, so that the expected number of tests in the new protocol is less than in the original one when $p > (3-\sqrt{5})/2$. This will show that a protocol satisfying 1-4 which contains a group-test of more than 1 element cannot be optimum.

Let B be the node on our test tree such that the group at B, which we call G has m elements where $m \geq 2$, but all tests beneath B are individual tests. We denote the branch of the test plan to be followed in case G is non-defective by B_1 and in case G is defective by B_2. Let a denote any element of group G. The new test protocol is as follows. Instead of testing G when we reach the node B, we test G-{a}. If defective, we know in particular that G is also defective and we can continue the test in the same manner as when G was found defective in the old plan. The additional information now available may enable us to infer the result of some tests (actually, only one); these of course will not be performed under the new plan. This is why we denote the branch by B_2' instead of B_2. If G-{a} is non-defective, we test a. If a is also non-defective, we have exactly the same information as when G turned out non-defective in the old plan, and we continue our testing in the same manner. When a turns out defective, we again proceed as in the case G was defective in the old plan, skipping tests the results of which can be inferred. This modification of B is denoted by B_2'' (see Figure 3.3.4.2). The remainder of the test protocol, i.e. everything which is not below B, is left unchanged.

For any sample, the number of tests required under the new plan is clearly at most one more than under the old plan. We complete the proof by pairing off each sample for which one more test is required under the new protocol with two samples for which there is a saving of at least one test under the new protocol. We will get a lower bound for the expected number of tests saved under the new protocol which will be positive when $p > (3-\sqrt{5})/2$.

In order to proceed, we need to characterize the set of samples for which more tests are needed under the new protocol:

LEMMA 3.3.4.1 - The only samples for which more tests are needed under the new plan are the ones for which node B is reached and G is found to be non-defective under the old protocol.

PROOF OF LEMMA - If G-{a} is defective, we follow the same procedure as under the old protocol, except for possibly skipping a test which previously had to be preformed. If G-{a} is non-defective and a is defective, we reach B_2'' having performed one more test than when we

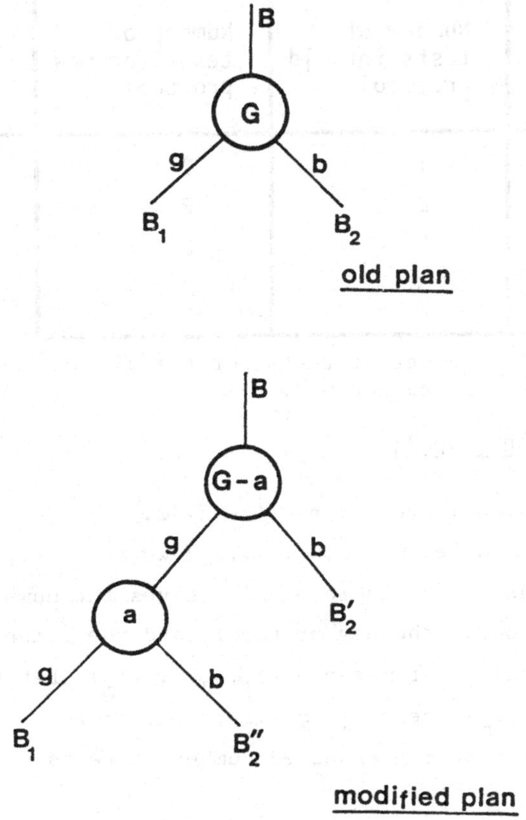

FIGURE 3.3.4.2 - Modification of a Group Testing Procedure containing Group-Tests of size greater than 1.

reached node B_2 under the old plan. However, under the old plan, the non-defectiveness of the (m-1) elements of G-{a} must be ascertained by (m-1) individual tests along the branch B_2 which can now be skipped. This proves the lemma.

We now distinguish between the cases m=2 and m>2. When m=2, let $G=\{s_1,s_2\}$ where s_1 represents element a; $s_i=1$ if the ith unit is defective and $s_i=0$ otherwise. Table 3.3.4.1 gives the number of tests needed under the old protocol, the new protocol, and the number of tests saved by all possible values of the pair (s_1,s_2). Let $P(i,j) = P(s_1=i,\ s_2=j)$ where i and j take values in the set {0,1}. Then we deduce from Table 3.3.4.1 that the expected number of tests saved is at least

(s_1, s_2)	Number of tests for old protocol	Number of tests for new protocol	Number of tests saved
0,0	1	2	-1
0,1	2	2	0
1,0	3	2	1
1,1	3	2	1

TABLE 3.3.4.1 - Number of tests for the old and new plan and the number of tests saved for m=2.

$$P(0,0)[-1+\frac{p}{q}+(\frac{p}{q})^2].$$

This expression is positive for $p > (3-\sqrt{5})/2$.

Now let $m \geq 3$, and let $G=\{s_1,s_2,\ldots,s_m\}$ where s_1 represents element a; s_i is defined as above. Table 3.3.4.2 gives the number of tests needed under the old protocol, the new protocol, and the number of tests saved for all possible values of m-tuple $(s_1,s_2,\ldots,s_m)$. Let $P(i_1,\ldots,i_m) = P(s_1 = i_1,\ldots,s_m = i_m)$ where i_j is either one or zero. Then we deduce from Table 3.3.4.2 that the expected number of tests saved is at least

$$p(0,0,\ldots,0)[-1+(m-1)\frac{p}{q}+(\frac{p}{q})^2].$$

This expression is positive for $p > (3-\sqrt{5})/2$, which proves the theorem.

Hence, TDMA is the optimum strategy for $p \geq (3-\sqrt{5})/2$. In the remainder of this section, we introduce a genie-aided argument which provides a lower bound to the performance of the optimum strategy for $0 \leq p \leq (3-\sqrt{5})/2$. Let $W^*_{SN}(M)$ represent the expected number of slots needed to process M users by using the optimum strategy with SN feedback.

THEOREM 3.3.4.3 - $W^*_{SN}(M) \geq H_g(M)$ where $H_g(M)$ is given by the following pair of recursive equations:

$$H_g(M) = 1 + \min_{1\leq x\leq M} \{q^x H_g(M-x) + (1-q^x)[F_g(x) + \sum_{i=1}^{x} \frac{q^{i-1}(1-q)}{1-q^x} H_g(M-i)]\} \qquad (3.3.4.1)$$

$$F_g(M) = 1 \text{ for } M \geq 2 \qquad (3.3.4.2)$$

$(s_1, s_2, \ldots \ldots \ldots, s_m)$	old plan	new plan	number of test saved
0 0 0 0 . . . . 0 0 0	1	2	-1
0 0 0 0 . . . . 0 0 1	m	m	0
0 0 0 0 . . . . 0 1 0	m+1	m+1	0
0 0 0 0 . . . . 1 0 0	m+1	m+1	0
.	.	.	.
.	.	.	.
.	.	.	.
.	.	.	.
.	.	.	.
0	.	.	.
0 0 1 0 0 . . . . 0 0 0	m+1	m+1	0
0 1 0 0 0 . . . . 0 0 0	m+1	m+1	0
1 0 0 0 0 . . . . 0 0 0	m+1	2	m-1
1 0 0 0 . . . . 0 0 1	m+1	m	1
1 0 0 0 . . . . 0 1 0	m+1	m+1	0
1 0 0 0 . . . . 1 0 0	m+1	m+1	0
.	.	.	.
.	.	.	.
.	.	.	.
.	.	.	.
.	.	.	.
.	.	.	.
.	.	.	.
.	.	.	.
.	.	.	.
1	.	.	.
1 0 1 0 . . . . 0 0 0	m+1	m+1	0
1 1 0 0 . . . . 0 0 0	m+1	m+1	0
the rest	m+1	m+1	0

TABLE 3.3.4.2

with the boundary conditions

$$H_g(0) = 0 \tag{3.3.4.3}$$

$$F_g(1) = 1 \tag{3.3.4.4}$$

PROOF - Assume procedure A3 is being used. Suppose that in the event that the feedback random variable has taken the value S for a slot in a given CRI, a beneficial genie provides us the index, i_1, for the least-index active user. There may or may not be more active users with higher indices than i_1; the genie does not provide us with any information about this matter. Hence after the genie's information has been provided, the remaining users with higher indices than i_1 who correspond to the slot which generated the S feedback, form a binomial set. Therefore, after transmitting the packet generated by user i_1 in the next slot, we investigate a new group of users according to the procedure A3. In the event of another S feedback, the genie will help us again. That is, whenever we are faced with an F-situaiion, the genie will help us. Doing so results in reaching an H-situation in only one slot after an F-situation. It is "clear" (see [22]) that the above procedure makes optimum use of the information provided by the genie. Therefore, the average number of slots needed to process M users using this procedure serves as a lower bound to optimum performance $W^*_{SN}(M)$ attainable without the genie's help. The performance of genie-aided protocol is simply given by (3.3.3.3)-(3.3.3.6) of the previous section, except that (3.3.3.4) is replaced by (3.3.4.2).

The values of $W_g(M)$ for M=64 are illustrated in Figure 3.3.4.3 as a function of p. Values of W(M) for procedure A3 also are shown in Figure 3.3.4.3. Note that the genie-aided procedure adapts itself to TDMA at p = 0.5. The following theorem shows that this value of p is independent of the number of users.

THEOREM 3.3.4.3 - If $p > 0.5$ and $n>0$, then $H_g(n) = n$ and $x(G_g(n)) = 1$

PROOF - The proof is by induction on n. Clearly $H_g(1) = 1$ and $x(H_g(k)) = 1$. Suppose the hypothesis of the theorem is true for $H_g(1)$, $H_g(2),\dots,H_g(n-1)$, and $x(H_g(1))$, $x(H_g(2)),\dots,x(H_g(n-1))$. Let 1(B) be the indicator function of event B. Then,

$$H_g(n) = 1 + \min_{1\le x\le n} \{q^x H(n-x) + (1-q^x)[1(x\neq 1) + \sum_{i=1}^{x} \frac{pq^{i-1}}{1-q^x} H(n-i)]\}$$

$$= 1 + \min\{q^x(n-x) + (1-q^x)1(x\neq 1) + \sum_{i=1}^{x} pq^{i-1}(n-i)\}$$

$$= 1 + \min\{q^x(n-x) + (1-q^x)1(x\neq 1) + np \sum_{1}^{x} q^{i-1} - p\sum_{1}^{x} iq^{i-1}\}$$

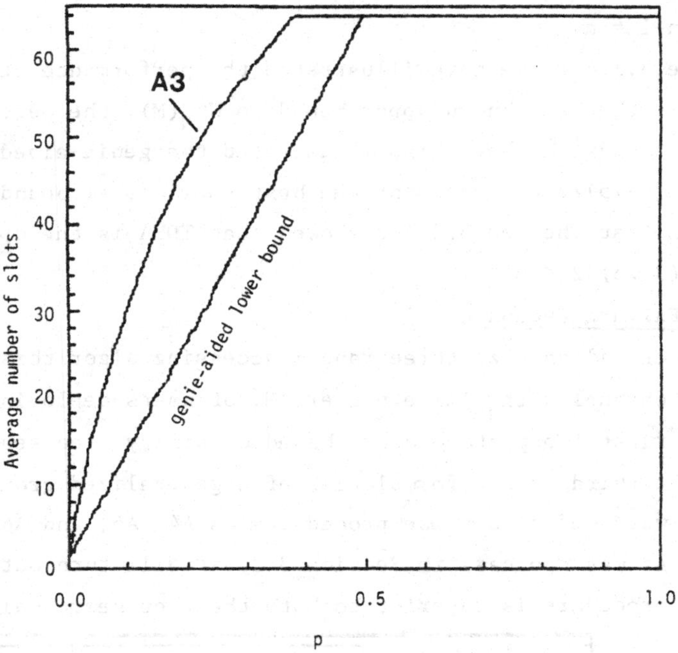

FIGURE 3.3.4.3 - Performance of the protocol A3 and the genie-aided lower bound for M=64.

$$= 1 + \min\{nq^x - xq^x + (1-q^x)1(x\neq 1) + np\,\frac{1-q^x}{1-q} - p\,\frac{1-(1+x)q^x}{(1-q)^2} - p\,\frac{q(1-q^x)}{(1-q)^2}\}$$

$$= 1 + \min\{nq^x - xq^x + (1-q^x)1(x\neq 1) + n - nq^x - 1 + (1+x)q^x - \frac{q(1-q^x)}{p}\}$$

$$= 1 + n + \min\{(1-q^x)1(x\neq 1) - 1 + q^x - \frac{q(1-q^x)}{p}\}$$

$$= 1 + n + \min_{1\le x\le n}\{(1-q^x)1(x\neq 1) - \frac{1-q^x}{p}\}$$

$$= 1 + n - \max_{1\le x\le n}\{\frac{1-q^x}{p} - (1-q^x)1(x\neq 1)\}$$

Let $g(x) = \frac{1-q^x}{p} - (1-q^x)1(x\neq 1)$

then $g(1) = 1$

and $g(x) = \frac{1-q^x}{p} - (1-q^x) = \frac{q(1-q^x)}{p}$ for $x > 1$

< 1 for $p < .5$

$\Rightarrow$ maximum is achieved at x = 1

$\Rightarrow H_g(n) = 1+n-1 = n$ QED

In Figure 3.3.4.4, we have illustrated the performance of procedure A3 to represent the best known upper bound to $W^*_{SN}(M)$, the performance of the optimum strategy. We have also illustrated the genie-aided procedure for $0<p<p^*_{SN} = (3-\sqrt{5})/2$ to represent the best known lower bound to $W^*_{SN}(M)$. Recall that Theorem 3.3.4.2 showed that TDMA is the optimum strategy for $(3-\sqrt{5})/2 \leq p \leq 1$.

3.4 Success/Failure Feedback

We consider and analyze three random accessing algorithms for a random access channel with finite number, M, of users deploying a SF feedback. The first two protocols are based on binary tree search methods and the third is a reformulation of a generalized group testing procedure. We will refer to these procedures as A4, A5, and A6, respectively. As was the case in Section 3.3, it will turn out that the group testing procedure is superior to both the tree search algorithms.

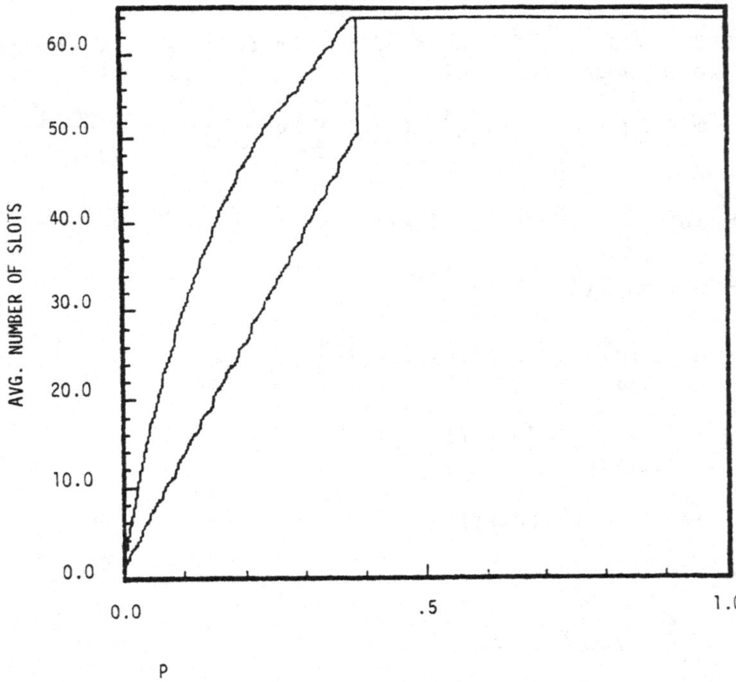

FIGURE 3.3.4.4 - Upper and lower bounds to the performance of the optimum strategy for SN feedback and M=64.

3.4.1 Procedure A4: Binary Tree Search Method. Consider a random access channel with $M=2^K$ users deploying a SF feedback, where channel and the users satisfy the model given in Section 3.1. Number the users $1,2,3,\ldots,2^K$ in the same fashion as Section 3.3.1. Consider the users as 2^K leaves of a binary tree.

PROCEDURE A4 - After an F feedback (if any), all transmitters on the top half of the tree retransmit in the very next slot; those on the bottom half transmit in the next slot after the confusion (if any) among those on the top half has been resolved. Continue in this manner until either there is no more confusion or you have reached the Kth level of the tree, at which point users will be transmitting their packets individually.

This protocol can be implemented by each of the users by growing the tree starting from the root node. (Simple procedures similar to the one introduced for protocol A1 can be easily given for this protocol.) The operation of this protocol can be best illustrated by an example. We apply the procedure A4 to the same example that we have used throughout. Therefore, assume M=8 and the initial conflict was among users number 3, 4, and 7. The corresponding binary tree is shown in Figure 3.4.1.1, and the feedback for each slot and the actions of each user is given in

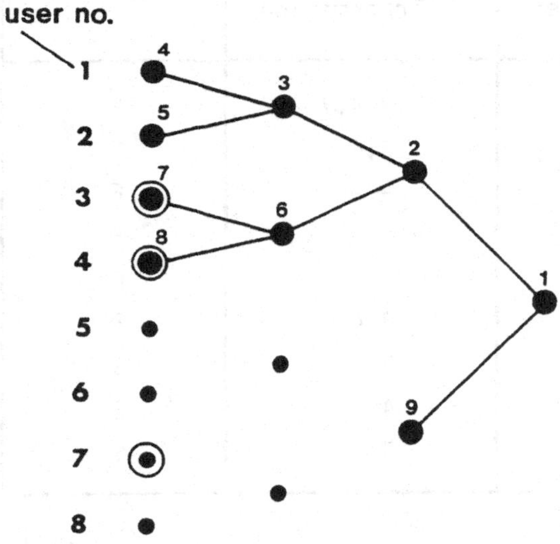

FIGURE 3.4.1.1 - Example for Procedure A4.

Table 3.4.1.1. In Figure 3.4.1.1, the numbers above each node indicate the index of the corresponding time slot in Table 3.4.1.1. We must now analyze this protocol. The definition of S(M), P(i,j), Y_{ij}, and n_{ij} are the same as those in Section 3.3.1.1. We seek the value of W(M) which is given by

$$W(M) = \sum_{i=0}^{K} \sum_{j=1}^{2^i} P(i,j) \tag{3.4.1.1}$$

any node n at level i is visited only if the transmission corresponding to the node at level (i-1), having node n as one of its two siblings, resulted in an F feedback. Hence

$$P(i,j) = 1 - 2^{K-i+1} pq^{2^{K-i+1}-1} \quad \text{for } i > 0 \tag{3.4.1.2}$$

$$= 1 \quad \text{for } i = 0$$

Using this expression for P(i,j) in Equation (3.4.1.1) we get

$$W(M) = 1 + \sum_{i=1}^{K} \sum_{j=1}^{2^i} [1 - 2^{K-i+1} pq^{2^{K-i+1}-1}]$$

Slot Number	Users Transmiting	Feedback
1	3,4,7	F
2	3,4	F
3	---	F
4	---	F
5	---	F
6	3,4	F
7	3	S
8	4	S
9	7	S

TABLE 3.4.1.1 - Channel Feedback and the action of users for example of Fig. 3.4.1.1.

$$= 1 + \sum_{i=1}^{K} 2^i[1 - 2^{K-i+1}pq^{2^{K-i+1}-1}] \qquad (3.4.1.3)$$

The values of W(M) for M=64 users have been illustrated in Figure 3.4.1.2 as a function of p. For comparison purposes, the line W(M)=M, which represents the performance of TDMA, also is shown in Figure 3.4.1.2. Note that protocol A4 is inferior to TDMA for all values of p. In particular note that W(M)=2M-1 for p=0. This is because, when p=0, nobody has a message to send yet the algorithm has to visit all of the nodes of the tree in order to resolve the confusion.

3.4.2 Procedure A5: Improved Binary Tree Search Method. The inefficient way that protocol A4 resolves its confusion about the packet transmissions when p=0 suggests the following modification. Introduce an (M+1)st user whom we shall refer to as the "auxiliary" user. The auxiliary user transmits one packet (any information packet will do) only in the first slot of every CRI. Therefore, when none of the users has a message to transmit, the auxiliary user will be the only one who

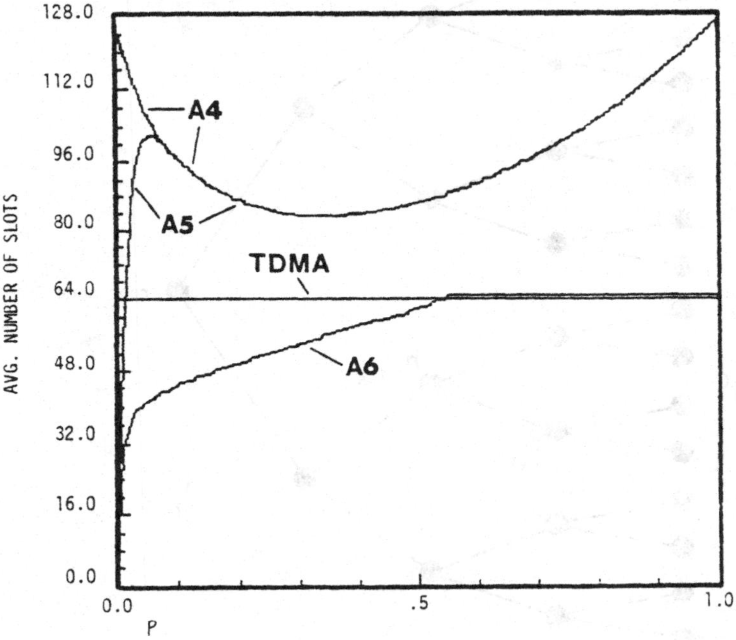

FIGURE 3.4.1.2 - Performance of procedures A4, A5, A6, and TDMA for SF feedback and M=64.

will be transmitting. This will result in an S feedback. This in turn will indicate to the protocol that none of the users have a message to send, where upon it need not visit all of the nodes of the tree in order to resolve its confusion. Hence, the auxiliary user transforms the SF feedback to an SN feedback for the first slot of each CRI.

PROCEDURE A5 - Same as procedure A4, except that the feedback from the first slot of every CRI should be interpreted as a SN feedback and used accordingly.

In order to analyze this protocol we need to express the probabilities P(i,j) in terms of p. The root node is always visited, so P(0,1)=1. The nodes at level one are visited only if at least one of the M users has a message to send, and hence $P(1,1) = P(1,2) = 1-q^{2^K}$. Values of P(i,j) for i>1 are computed with the aid of the binary tree of Figure 3.4.2.1. Let S_1 and S_2 represent the number of active users in the sets

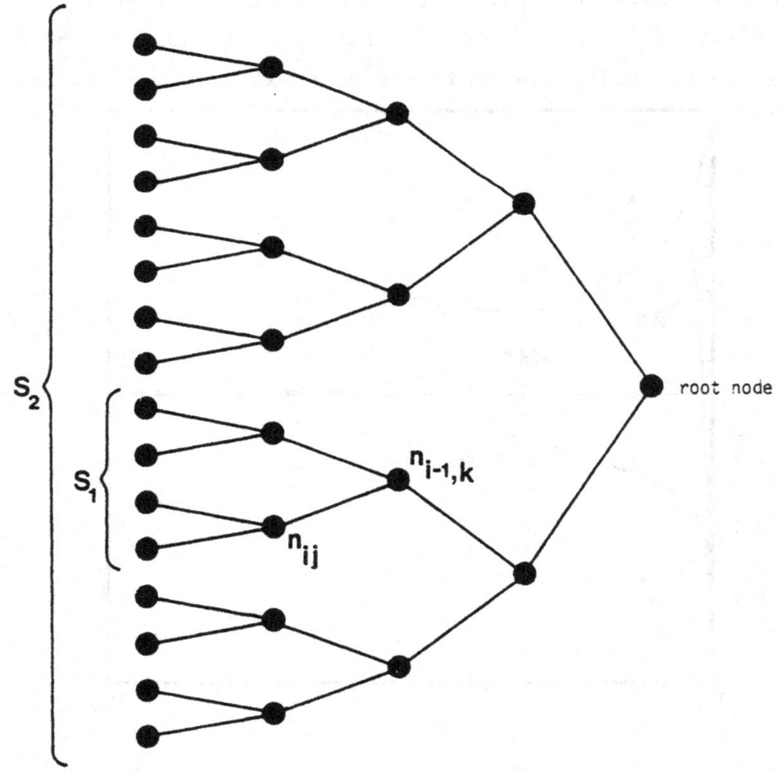

FIGURE 3.4.2.1

indicated in Figure 3.4.2.1. Let $S_3 = S_2 - S_1$. We then deduce that for $i>1$:

$$
\begin{aligned}
P(i,j) &= P(\text{visiting node } n_{ij}) \\
&= P(S_1 \neq 1 \text{ and } S_2 \geq 1) \\
&= P(S_1 = 0,\ S_2 \geq 1) + P(S_1 \geq 2,\ S_2 \geq 1) \\
&= P(S_1 = 0)P(S_3 \geq 1) + P(S_1 \geq 2) \\
&= [q^{2^{K-i+1}}][1-q^{2^K - 2^{K-i+1}}] + [1-q^{2^{K-i+1}} - 2^{K-i+1}pq^{2^{K-i+1}-1}]
\end{aligned}
\tag{3.4.2.1}
$$

Therefore,

$$
W(M) = 1 + 2[1-q^{2^K}] + \sum_{i=2}^{K} 2^i P(i,j) \tag{3.4.2.2}
$$

where $P(i,j)$ is given by Equation (3.4.2.1). Values of $W(M)$ for $M=64$ are shown in Figure 3.4.1.2 as a function of p, along with the corresponding values for procedure A6 and TDMA. Note that protocol A5 improves A4 only for very small values of p as it was designed to do.

<u>3.4.3 Procedure A6: Generalized Group Testing Method.</u> In the last two sections we studied the performances of two CRAs which were based on binary tree search methods. The first protocol, A4, was inferior to TDMA for all values of p and the next one, A5, improved the performance of A4 for very small values of p by deploying an "auxiliary user." We shall now use this auxiliary user with a group testing method to describe procedure A6.

Let a=b=1 in the Definition 3.2.1 for a generalized group testing problem. The resulting test, $T(1,1;G)$, corresponds to a SF feedback in our communication problem. Let n represent the number of unclassified users at the beginning of some slot. We will refer to the beginning of the first slot of every CRI, where all the M users are in a binomial set, as an H-situation. If at the beginning of some later slot all of the users are in a binomial set, we call that an H-situation. Let $H(M)$ be the average number of slots needed to resolve an H-situation, and let $\tilde{H}(n)$ be the average number of slots needed to resolve an H-situation when the binomial set is of size n. Also assume that we have an auxiliary user similar to the one described in Section 3.4.2. Let $x(H)$ and $\tilde{x}(H)$

be integers to be optimized later. The procedure A6 can now be described as follows:

PROCEDURE A6 - If we have an H-situation, the x(H) users from the binomial set with the least indices are enabled; If this results in an F feedback, the same x(H) users are enabled individually in the next x(H) slots, after which the size of the binomial set is reduced by x(H). If we have an $\tilde{H}$-situation, then we do as above except using $x(\tilde{H})$ instead of x(H).

The role of the auxiliary user becomes clear by studying the pair of recursive equations describing the operation of procedure A6. These equations are:

$$H(M) = 1 + \min_{1\leq x\leq M} \{q^x \tilde{H}(M-x) + (1-q^x)[x + \tilde{H}(M-x)]\} \qquad (3.4.3.1)$$

$$\tilde{H}(n) = 1 + \min_{1\leq x\leq n} \{xpq^{x-1} \tilde{H}(n-x) + (1-xpq^{x-1})[x + \tilde{H}(n-x)]\} \qquad (3.4.3.2)$$

with the boundary condition

$$H(0) = \tilde{H}(0) = 0 \qquad (3.4.3.3)$$

Equation (3.4.3.1) represents the operation of protocol for the very first slot of any CRI where the auxiliary user is present. The next equation is for the rest of the slots where the auxiliary user is not present.

We have applied the protocol to our usual example and the results are shown in Table 3.4.3.2. The optimum values of x(H(M)) and $x(\tilde{H}(n))$ for M=8 and p=0.15 are shown in Table 3.4.3.1. The performance of procedure A6, which is represented by values of W(M) = H(M), is shown in Figure 3.4.1.2 as a function of p. For comparison purposes, the corresponding values of W(M) for procedures A4, A5, and TDMA also are shown in Figure 3.4.1.2. Note that procedure A6 outperforms A4 and A5 for all values of p, performs better than TDMA for small values of p, and performs almost as well as TDMA for higher values of p. Let $p^*_{SN}(M)$ be the value of p where procedure A6 enables each one of the users individually. The value of H(M) for procedure A6 is greater than that of TDMA by 1 for $p>p^*_{SF}(M)$. The reason for this is the one slot used

$x(\tilde{H}(8))$	8
$x(\tilde{H}(7))$	7
$x(\tilde{H}(6))$	6
$x(\tilde{H}(5))$	5
$x(\tilde{H}(4))$	4
$x(\tilde{H}(3))$	3
$x(\tilde{H}(2))$	2
$x(\tilde{H}(1))$	1
$x(H(8))$	2

TABLE 3.4.3.1 - Optimum values of $x(H(M))$ and $x(\tilde{H}(n))$ for M=8 and p = .15.

Slot Number	Situation	x	Users Transmiting	Feedback
1	$H(8)$	2	auxiliary	S
2	$\tilde{H}(6)$	6	3,4,7	F
3			3	S
4			4	S
5			---	F
6			---	F
7			7	S
8			---	F

TABLE 3.4.3.2 - Operation of A6 for M=8, p = .15 when users #3, 4, and 7 have messages to send.

by the auxiliary user. Hence, a better strategy would be to use A6 for $0 < p < p^*_{SF}(M)$ and then use TDMA for $p \geq p^*_{SF}(M)$. In contrast to procedure A3, $p^*_{SF}(M)$ does depend on the value of M; it is believed to be an increasing function of M.

3.4.4 On the Performance of the Optimum Strategy for SF Feedback.

The performance of procedure A6 for $0 < p < p^*_{SF}(M)$ and that of TDMA for $p > p^*_{SF}(M)$ provide us with upper bounds to $W^*_{SF}(M)$, the performance of the optimum strategy for SF feedback. We now seek a lower bound to $W^*_{SF}(M)$.

Molle [22] has shown that for a system with 0,1,e-ternary feedback and an infinite horizon Bernoulli-user model, the efficiency of the optimum protocol is bounded above by a function $\rho(p)$ which is given in [22] and has the property that $\rho(1) = 1$ and $\lim_{p\to\infty}\rho(p) = 0.6731$. Clearly the optimum protocol for the 0,1,e feedback will perform at least as well as the optimum protocol for any binary feedback. Moreover, the performance of the optimum protocol for the finite-user model can only be improved by considering an infinite horizon. Therefore, since efficiency is defined as the average fraction of slots containing exactly one packet, we get

$$\rho(p) \geq \frac{pM}{W^*_{SN}(M)}$$

or

$$W^*_{SN}(M) \geq \frac{pM}{\rho(p)}$$

Using the values of $\rho(p)$ given in [22], the above bound is illustrated in Figure 3.4.4.1 for M=64. The values of W(M) for protocol A6 for $0 \leq p \leq p^*_{SN}(M)$ and the TDMA for $p \leq p^*_{SN}(M)$ are also shown in Figure 3.4.4.1 to represent an upper bound to $W^*_{SF}(M)$.

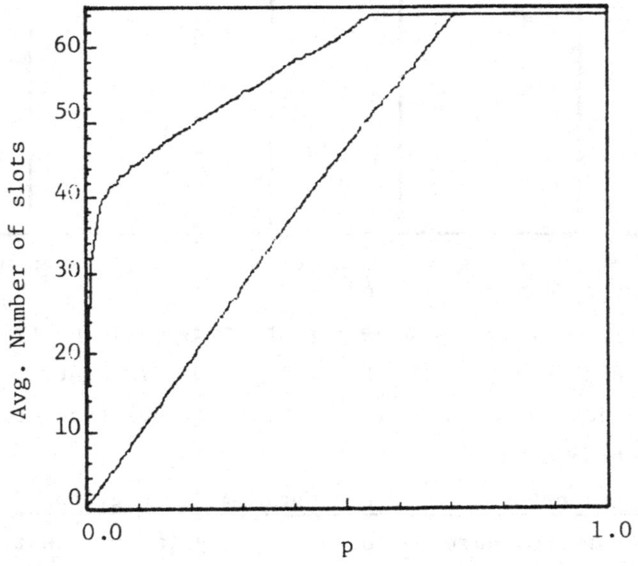

FIGURE 3.4.4.1 - Upper and lower bound to the performance of the optimum strategy for SF feedback and M=64.

REFERENCES

[1] J. H. Wittman, "Categorization of Multiple-Access/Random Access Modulation techniques," IEEE Trans. Commun. Tech., Vol. COM-15, No. 5, pp. 724-725, Oct. 1967.

[2] J. W. Schwartz, J. M. Aein, and J. Kaiser, "Modulation Techniques for Multiple Access to a Hard-Limiting Satellite Repeater," Proc. IEEE, Vol. 54, pp. 763-777, May 1966.

[3] S. S. Lam, "A New Measure of Characterizing Data Traffic," IEEE Trans. Commun., Vol. COM-26, No. 1, pp 137-140, Jan. 1978.

[4] N. Abramson, "The ALOHA System - Another Alternative for Computer Communications," AFIPS Conf. Proc., Vol. 37, AFIPS Press, pp. 281-285, 1970.

[5] L. G. Roberts, "ALOHA Packet System with and without Slots and Capture," ARPANET Satellite System Note 8, ARPA Network Info. Ctr., Stanford Res. Inst., Stanford, CA, June 1972. Reprinted in Computer Communications Review, Vol. 5, April 1975.

[6] L. Kleinrock, Queueing Systems, Vol. 2: Computer Applications, Wiley, NY, 1976.

[7] G. Fayolle, E. Gelenbe, and J. Labetoulle, "Stability and Optimal Control of the Packet Switching Broadcast Channel," J. of the Assoc. of Comp. Mach., Vol. 24, No. 3, pp. 375-386, July 1977.

[8] M. J. Ferguson, "On the Control, Stability, and Waiting time in a Slotted ALOHA Random-Access System", IEEE Trans. Commun., Vol. COM-23, No. 11, pp. 1306-1311, Nov. 1975.

[9] S. S. Lam, and L. Kleinrock, "Packet Switching in a Multiaccess Broadcast Channel: Dynamic Control Procedures," IEEE Trans. Commun., Vol. COM-23, No. 9, pp. 91-94, Sept. 1975.

[10] V. A. Mihailov, "A Random Access Algorithm for a Broadcast Channel," presented in the 5th Int'l. Symposium on Info. Theory, Tbilisi, Georgia, USSR, Vol. 3, pp. 83-86, in Russian, 3-7, July, 1979.

[11] B. Hajek, and T. Van Loon, "Decentralized Dynamic Control of a Multi-Access Broadcast Channel," to appear in IEEE Trans. Auto. Control, 1982.

[12] J. I. Capetanakis, "The Multiple Access Broadcast Channel: Protocol and Capacity Consideration," Technical Report ESL-R-806, MIT, Cambridge, MA, March 1978.

[13] J. I. Capetanakis, "Tree Algorithm for Packet Broadcast Channels," IEEE Trans. Info. Theory, Vol. IT-25, No. 5, pp. 505-515, Sept. 1979.

[14] J. I. Capetanakis, "Generalized TDMA: The Multi-Accessing Tree Protocol," IEEE Trans. Commun., Vol. COM-27, No. 10, pp. 1476-1484, Oct. 1979.

[15] B. S. Tsybakov, and V. A. Mikhailov, "Slotted Multi-Access Packet-Broadcasting Feedback Channel," in Russian, Problemy Peredachi Informatsii, Vol. 14, No. 4, pp. 32-59, Oct.-Dec. 1978. Translated as "Free Synchronous Packet Access in Broadcast Channel with Feedback," Problems of Information Transmission, Vol. 14, No. 4, pp. 259-280, Plenum Publishing Co., NY, 1978.

[16] R. G. Gallager, "Conflict Resolution in Random Access Broadcast Networks," Proc. of the AFOSR Workshop in Communications Theory and Applications, Provincetown, MA, pp. 74-76, 17-20, Sept., 1978.

[17] J. Mosely, "An Efficient Contention Resolution Algorithm for Multiple Access Chanels," Technical Report LIDS-TH-918, MIT, Cambridge, MA, June 1979.

[18] J. L. Massey, "Collision-Resolution Algorithms and Random Access Communications," in Multi-User Communications, Edited by G. Longo, Springer-Verlag, CISM Courses and Lecture Series, No. 265, 1981.

[19] T. Berger, "The Poisson Multiple Access Conflict Resolution Problem," in Multi-User Communications, Edited by G. Longo, Springer-Verlag, CISM Courses and Lecture Series, No. 265, 1981.

[20] N. Mehravari, "The Straddle Algorithm for Conflict Resolution in Multiple Access Channels," MS Thesis, School of Elect. Engr., Cornell Univ., Ithaca, NY, May 1980.

[21] N. Pippenger, "Bounds on the Performance of Protocols for a Multiple Access Broadcast Channel," IEEE Trans. Info. Theory, Vol. IT-27, No. 2, pp. 145-152, March 1981.

[22] M. L. Molle, "On the Capacity of the Infinite Population Multiple-Access Protocols, IEEE Trans. on Info. Theory, Vol. IT-28, No. 3, 396-402, May 1982.

[23] R. Cruz, and B. Hajek, "A New Upper-Bound to the Throughput of a Multi-Access Broadcast Channel," IEEE Trans. on Info. Theory, Vol. IT-28, No. 3, pp. 402-406, May 1982.

[24] B. Hajek, "Information of Partitions with Applications to Random Access Communications," Coordinated Sciences Lab. and Dept. of Elect. Engr., Univ. of Ill., Urbana, Ill., June 1980, (To appear in IEEE Trans. on Info. Theory, September, 1982.)

[25] P. Humblet, private communication between J. Massey and P. Humblet, mentioned in [18].

[26] V. A. Mihailov, and B. S. Tsybakov, "An Upper Bound to Capacity of Random Multiple Access System," In Russian, Problemy Peredachi Informatsii, Vol. 17, No. 1, pp. 90-95, Jan.-Mar. 1981, Moscow, USSR. Translated as "Upperbound for the Capacity of Random Multiple Access System," Problems of Information Transmission, Vol. 17, No. 1, pp. 63-67, Plenum Publishing Co., NY, 1981.

[27] T. Berger, N. Mehravari, and G. Munson, "On Genie-Aided Upper Bounds to Multiple Access Contention Resolution Efficiency," Proc. of 1981 Conf. on Info. Sciences and Systems, John Hopkins Univ., Baltimore, MD, 25-27, March 1981.

[28] B. S. Tsybakov, "Resolution of Conflict with Known Multiplicity," in Russian, Problemy Peridachi Informatsii, Vol. 16, No. 2, pp. 69-82, Apr.-June 1980, Moscow, USSR. Translated as "Resolution of a Conflict of Known Multiplicity," Problems of Information Transmission, Vol. 16, No. 2, Plenum Publishing Co., NY, 1980.

[29] L. Georgiadis, and P. Papantoni-Kazakos, "A Collision Resolution Protocol for Random Access Channels with Energy Detectors," EECS Dept., Univ. of Conn., Storrs, CT, 1981. Submitted to IEEE Trans. on Commun.

[30] T. Berger and N. Mehravari, "Conflict Resolution Protocols for Random Multiple Access Channel with Binary Feedback" Proc. of Nineteenth Annual Allerton Conf. on Commun., Control, and Computing, Sept. 30-Oct. 2, 1981.

[31] D. F. Towsley, and J. K. Wolf, "An Application of Group Testing to the Design of Multi-User Access Protocols," Dept. of EE and Computer Engr., Univ. of Mass., Amherst, MA. Presented at the Nineteenth Annual Allerton Conf. on Commun., Control and Computing, Sept. 30-Oct. 2, 1981.

[32] Robert Dorfman, "Detection of Defective Members of Large Populations," Ann. Math. Stat., Vol. 14, pp. 436-440, 1943.

[33] Andrew Sterrett, "On the Detection of Defective Members of Large Populations," Ann. of Math. Stat., Vol. 28, pp. 1033-1036, 1957.

[34] Milton Sobel, and Phyllis A. Groll, "Group Testing to Eliminate Efficiently all Defectives in Binomial Sample," BSTJ, Vol. 38, pp. 1179-1253, Sept. 1959.

[35] Peter Ungar, "The Cutoff Point for Group Testing," Commun. on Pure and Appl. Math., Vol. 13, pp 49-54, 1960.

[36] J. F. Hayes, "An adaptive Technique for Local Distribution," IEEE Trans. on Commun., Vol. COM-26, No. 8, pp. 1178-1186, Aug. 1978.

[37] F. C. Schoute, "Decentralized Control in Packet Switched Sattelite," IEEE Trans. on Auto Control, Vol. AC-23, No. 2, pp. 362-371, No. 5, pp. 794-796, Oct. 1979.

[38] P. Varaiya, and J. Walrand, "Decentralized Control in Packet Switched Sattelite," IEEE Trans. on Auto. Control, Vol. AC-24, No. 5, pp. 794-796, Oct. 1979.

[39] T. Berger, "Studies in Multiterminal Communication, Information and Decision Theory," Research proposal submitted to the National Science Foundation, School of Electrical Engineering, Cornell Univ., Ithaca, NY, June 1979.

[40] N. Pippenger, Private Communication between N. Pippenger and T. Berger, February 1979.

[41] N. Mehravari, Random Multiple-Access Communication with Binary Feedback, Ph.D. Thesis, School of Elect. Engr., Cornell Univ. August, 1982.

SECURITY IN DISTRIBUTED MOBILE-USER RADIO NETWORKS

Anthony Ephremides
Electrical Engineering Department
University of Maryland
College Park, MD 20742

(Based on lectures delivered in the Laplace Session of the CISM Summer School on the topic of Secure Digital Communications, Udine, Italy, June, 1982.)

Preface

I wish to thank Professor Giuseppe Longo for giving me the opportunity to present the material that follows in the 1982 CISM Summer School on Secure Digital Communications. My emphasis is placed on issues most pertinent to Packet-Radio Communications and especially to mobile multi-user environments that are controlled in a distributed (non-central) fashion. Much of the material in these lectures is new and unpublished, and represents recent work by myself and my colleagues.

Lecture 1: Introduction

The basic threats to any form of communication are the following:

a. Interference

b. Interception

c. Falsification

Each of these threats may appear in a variety of ways depending on the communication environment as well as on the physical one. The means for combatting these threats are therefore also dependent on the specific application conditions. In Fig. 1 (a,b,c) we show in simplified schematic form the essential distinction among these threats.

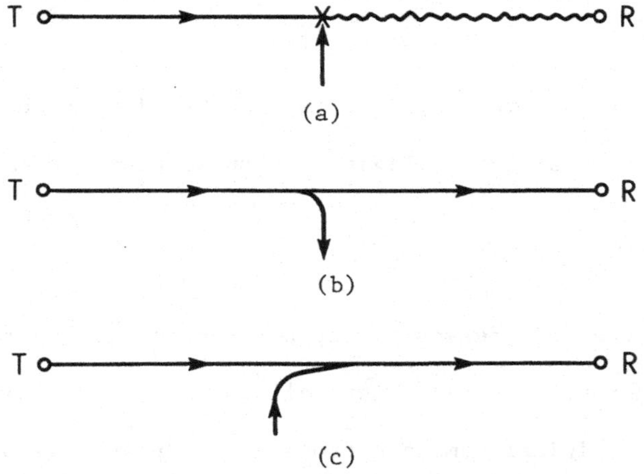

Fig. 1

In the case of interference we mean the disabling of a communication link by non-physical means. In the case of radio communications this threat is commonly referred to as jamming.

In the case of interception we mean the intrusion of an unauthorized receiver capable of interpreting in some way the transmitted messages.

A limited form of such interpretation may consist of identifying or locating a specific transmitter; this case is known in radio communications as direction finding.

Finally, in the case of falsification we mean the intrusion of an unauthorized transmitter (imposter) capable of sending interpretable messages through the link (spoofing).

There are three general approaches to combatting these threats. One is cryptography which operates at the "software" level and uses various techniques for encoding (encrypting) the transmitted messages. Cryptography is best suited to preventing interception and in allowing authentication. Another approach is the use of spread spectrum techniques. This approach operates at the waveform design level and includes "software" (coding) schemes as well as "modem" (modulation) methods. This approach is best suited to countering the interference threat. The final approach is physical protection which operates essentially at the "hardware" level and has special meaning only in the case of radio communications where appropriate antenna configurations and waveform polarization can achieve high degrees of directivity and is of course suited to countering up to some level all major threats in their radio forms.

Our purpose here is not to discuss cryptography or antenna arraying techniques. We do not wish either to merely examine the known methods of spectrum spreading such as frequency hopping, direct sequence, or hybrid techniques and evaluate their performance. Instead we intend to focus on radio communications and examine special issues concerning the described threats and their remedies and identify novel problems that arise in connection with these threats when there are many users who share the communication medium, who are free to move about, and who must cooperate in sharing the common channel as well as in countering the various threats.

However in doing so we won't simply consider the classical problems of multiple access that naturally arise in this context. Instead we will try to study the effects on multiple access capability and performance that the need of protection against the interference threats imposes on such distributed mobile-user radio network systems.

We begin by reviewing briefly the main approaches taken in addressing the basic multiple access problem. The traditional way of allocating bandwidth is of course via frequency division. Each user is provided with a different portion of the entire channel resource. Each portion is dedicated to a single user for his exclusive use and is either permanently assigned or belongs to that user until reallocation is made according to some demand assignment scheme. This method has been known as frequency division multiple access (FDMA).

What is wrong with frequency division? Consider the case of five users as in Figure 2. Suppose they are all within radio communication range of each other. Under frequency division there should be a total of twenty one-way channels, or ten full-duplex links. The intermodulation at each node and the mutual interference would be at substantial levels. Furthermore, the link management technique, in case of demand-assignment, would require considerable sophistication particularly if it is to be implemented in a distributed fashion and if not all users are within single hop range of each other. In the latter case, even for non-demand assignment cases, the problem of efficient frequency re-use has been a major one, particularly for mobile users, and has received only recently some partially satisfactory solutions [1] .

The duality between time and frequency in communication theory makes the time division multiple access (TDMA) techniques conceptually apparent. Instead of partitioning the available bandwidth among the users, the time-of-use of the entire bandwidth can be equivalently partitioned, resulting in a capacity distribution identical to the one achieved by the FDMA method. Thus each user has exclusive rights on the use of the entire channel during predetermined, periodically recurring time slots. These time slots are dedicated to each user. The allocation of slots need not be permanently fixed, but can be managed in some dynamic way. Demand access operation in the time domain creates many more possibilities than in the frequency domain.

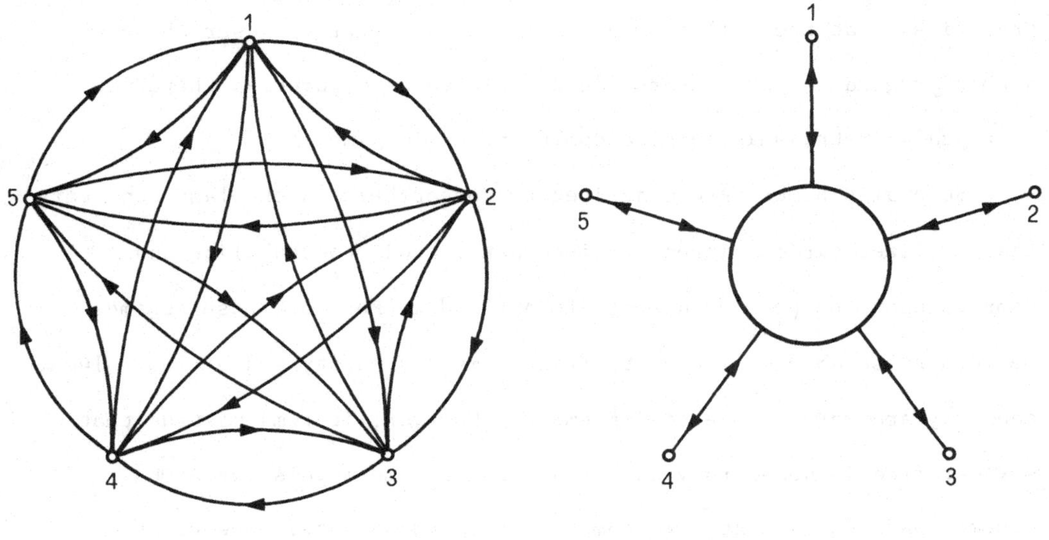

Fig. 2

Time division methods became attractive due to the elimination or reduction of radio interference. Consider again the setup of Figure 2 with the five users. Instead of ten full-duplex channels there is only a single channel. The price paid for this improvement is of course the need for synchronization. The main advantage of time-domain methods is the flexibility offered by the various ways that time slots can be used or assigned. One such way is the totally random use, often called the ALOHA scheme. Since the introduction of TDMA it had been observed that when some users had low duty cycle transmission or when the overall traffic level was generally low, many slots were not used and as a result the channel capacity was under-utilized. The ALOHA scheme [2] constituted a bold experimentation with an approach that was the extreme opposite of the orderly fixed allocation of slots used in TDMA. Furthermore, while still operating in the time-domain it did not require synchronization. It simply postulated that every user would attempt transmission at the full channel bandwidth whenever the user chose to, without regard to other users' needs. In cases of overlap there would be repeated retransmissions randomly spaced in time.

Several schemes were conceived by researchers in the last ten years that utilized various ingeneous feedback methods to stabilize and improve upon the performance of ALOHA [3-6] . Most of these schemes perform reasonably well when traffic is mostly bursty and of low volume, and they are very simple to implement. However, it remains true that when traffic is non-bursty and/or at peak volume levels the ALOHA scheme and its variants are completely inappropriate. The value of this fundamental and perhaps revolutionary departure from classical thinking in the allocation of resources lay primarily in that it opened the way for the careful and controlled introduction of contention schemes of multiple access.

In the time domain it is remarkably straight-forward to implement and manage demand access methods. By allocating a separate resource, called the control, or reservation channel, to narrow bandwidth reservation messages it is possible to make near perfect use of the main channel by allocating it as needed to the requesting users. The control may consist of a separate frequency channel or of a separate set of time slots entirely in the time domain. There are many reservation and polling schemes proposed in the literature and most of them are modifications of the same basic demand assignment scheme [3] . Where they may differ is the method specified for accessing the control channel. Obviously, reservations cannot be used to access the reservation channel. Thus, one must resort to other schemes for that purpose, such as TDMA, contention-based, or hybrid schemes. Furthermore, reservation techniques are not easy to implement when there is no central controller.

From the preceding discussion it follows that, unless the traffic patterns generated by the users of a multiple access system fall consistently into certain categories (bursty and low volume, or high duty cycle and high volume respectively) neither fixed allocation nor pure contention protocols are suitable.

Thus there is a need for protocols that combine the virtues of both extremes (namely of dedicated allocation and of sharing) and can adapt to either one as the traffic characteristics change. Although there is a wealth of imaginative schemes proposed and analyzed or simulated in the literature [4] , the quest for improved multi-access protocols remains a major component of basic research in the area of multi-user communications.

A method more akin to a pure contention protocol, but with a systematic (rather than purely random) retransmission strategy that has some reservation characteristics, was proposed by J. Capetanakis [7] . The Capetanakis conflict resolution, or tree, algorithm exists in several versions. It is not an exaggeration to say that his approach has provided as much of an infusion of fresh thought to multi-access research as did Abramson's conception of the ALOHA scheme over ten years ago.

As with the passage from frequency division to time division, it takes a simple, direct conceptual step to visualize the channel resource modeled as a slice of the frequency-time plane (Figure 3). Frequency division partitions this slice into strips parallel to the frequency axis. It is clear that the two-dimensional resource can be divided in other ways as well. Consider for example the mesh of Figure 3 where the bandwidth has been partitioned into M bands and the time axis has been slotted. User i can be assigned an arbitrary sequence of frequency bands, periodic or not, as illustrated in the figure. So long as every other user is assigned a sequence that does not overlap with that of any other user the resulting partition is a form of time-varying frequency division which has been called frequency hopping or code division since each sequence can be thought of as a code.

Although there are other ways of implementing code-division, such as pseudo-noise sequence modulation or hybrid forms, the frequency

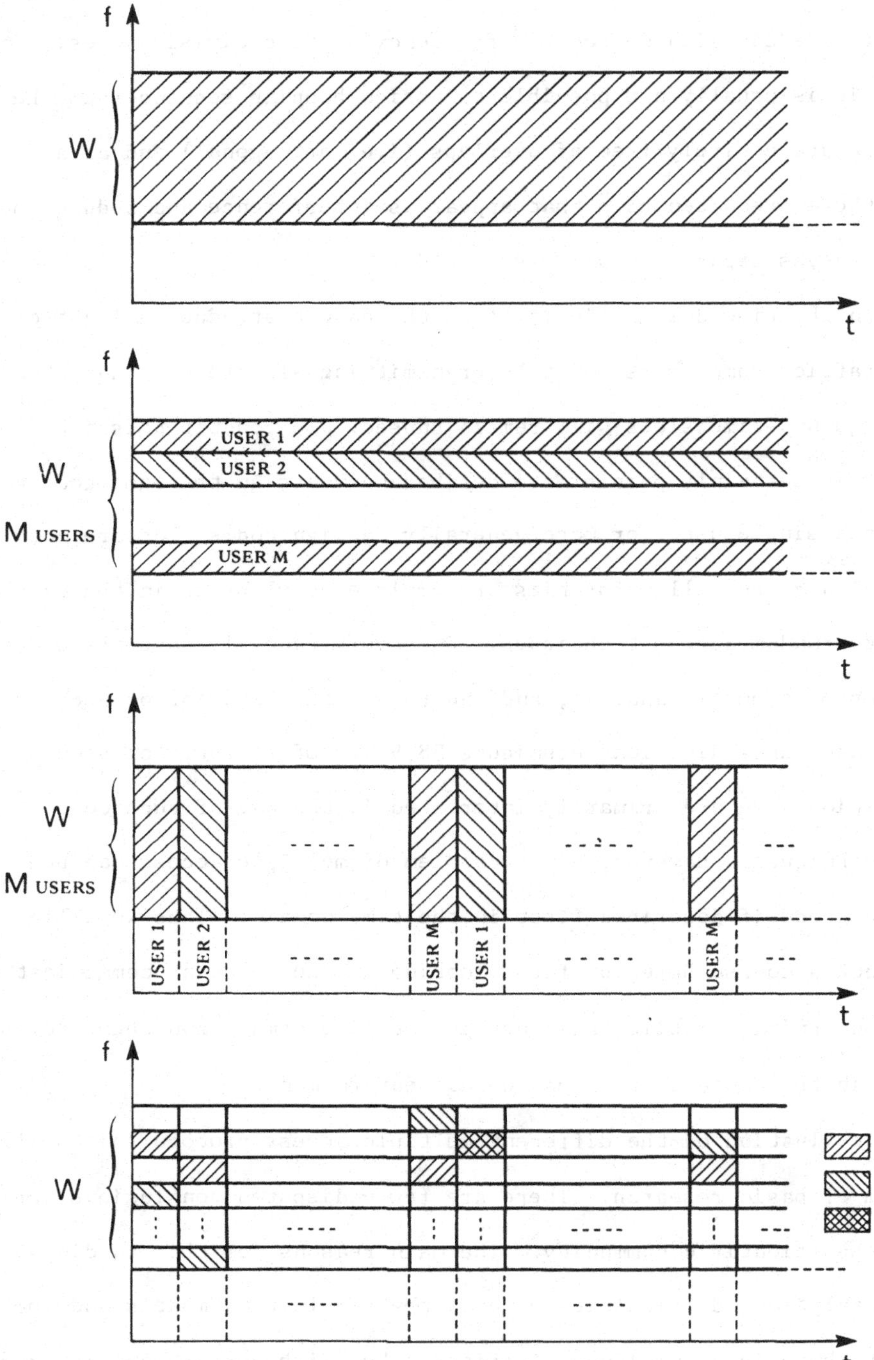
f
W
t
f
USER 1
USER 2
W
M USERS
USER M
t
f
W
M USERS
USER 1
USER 2
USER M
USER 1
USER M
t
f
W
USER 1
USER 2
"HIT"
t

Fig. 3

hopping method illustrates most lucidly the equivalence (conceptually at least) of code division methods to frequency or time division ones. In practice it is usually not possible to assign hopping sequences to the users that are entirely free of overlaps (i.e. orthogonal) and as a result, there are often hits that appear as interference and reduce the multiple access capability of this method.

As in the time domain, it is often the case that, due to light or bursty traffic, some users won't be transmitting all the time and thus their frequency-time slots will remain unused. To counter this inefficiency it is certainly possible to deliberately assign the same code to more than a single user, or more generally, assign codes that are not orthogonal and thus allow for hits or complete overlaps as in the case of ALOHA-type random access techniques. Only very recently have there been any attempts to model, analyze, and understand the behavior of such contention-based code division techniques [8,9]. Of course, for secure communications, we are primarily interested in the use of such code-domain techniques. However, every problem of multiple access can be best understood if described first in the time domain. Thus we will follow such a course here, by first looking at the relevant communication issues concerning a mobile radio net in the time domain and then addressing them in the context of a code domain environment.

The evaluation of the different multiple access protocols is still a problem of basic research. There are few undisputed conclusions agreed upon by the scientific community. The main reasons for this tardiness are the analytical difficulties of even the simplest of models and the fact that there is a multitude of criteria by which performance is judged.

In the next lectures we will address in some detail some of these issues.

Lecture 2: The Capture Phenomenon

In the first lecture we surveyed the main classes of multiple access protocols and we saw that in all cases a key assumption was that when two or more packets are transmitted simultaneously or even partially overlapping in time total and destructive interference occurs for all packets involved. Experimentation since the early days of the ALOHA studies, corroborated later by satellite tests, revealed an astonishing phenomenon. Often when two or more packets were transmitted simultaneously the receiver "captured" one of them and received it correctly while, of course, the other packets were rejected. Notice that we are not talking about the selective reception of which a receiver is capable if a spread spectrum modulation scheme is used. In fact in this lecture we will try to analyze the behavior of some spread spectrum schemes when "capture" occurs as a separate, additional phenomenon. Capture is not a "threat" to radio communication but is something that affects performance of other threats (such as interference) as well as performance of remedies to these threats (such as spectrum spreading) and thus deserves attention. Trying to understand and analyze this phenomenon is the object of this lecture.

First let us assume that spread spectrum signaling is not used. What physical explanations are there for the capture phenomenon? Consider that users are mobile and geographically dispersed. There is a number of factors that may cause capture. First of all antennae have directivity properties. Depending on their orientation high selectivity can be achieved at the receiver. Distance from the receiver is another factor. Transmitter power variations or differences may also affect reception. Additionally, propagation conditions vary with location, sea state, and time-of-day. Clearly they can affect the level of the received power of different signals differently. Finally lack of synchronization due to clock fluctuation as well as due to motion may produce timing differentials at the receiver that permit "phase-locking" to the signal with the early time advantage (i.e. a phenomenon analogous to the very basic characteristic of spread spectrum schemes).

Earlier attempts to model capture [10-12] focused on deterministic explanations that were based almost exclusively on distance; that is, it was assumed that in perfectly synchronized transmissions the user closer to the receiver had both timing and power advantages and thus would prevail. Consider for example Abramson's model in Fig. 4.

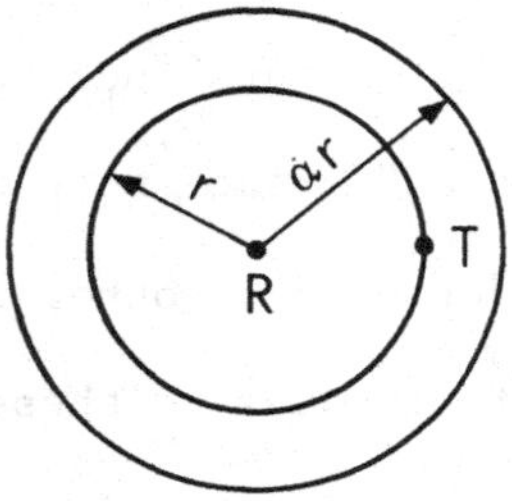

Fig. 4

User T is assumed to prevail at the receiver R whenever a user situated outside the bigger circle is transmitting at the same time with T. If a=1, capture is called "perfect". By assuming that G(r) and S(r) represent the densities of traffic rates for total channel traffic and for throughput to the receiver respectively and by assuming uniform distribution of sources around the circle, the quantities G(r)·2 r·dr and S(r)·2 r·dr represent the traffic rates due to transmission from sources located at a distance r from the receiver. Thus the overall channel traffic and throughput rates are given by

$$G = \int_0^\infty G(r)2\pi r dr \qquad S = \int_0^\infty S(r)2\pi r dr$$

Let us recall that for random access without capture the relationship between G and S is either $S=Ge^{-2G}$ for unslotted access with a corresponding maximum throughput rate of 1/2e packets per slot or $S=Ge^{-G}$ for slotted access with a corresponding maximum of $\frac{1}{e}$.

As expected, the assumption of capture produces an advantage. A simple analysis shows that the maximum throughput rate is equal to $\frac{1}{2a^2}$ for unslotted access and to $\frac{1}{a^2}$ for slotted (i.e. for perfect capture it is possible to sustain maximum utilization of the channel resource by achieving a throughput rate of 1 packet per slot). Another interesting result for this model is that for each sustainable throughput value S_0 there is a radius r_0 such that $G(r)= \infty$ for $r>r_0$. The critical value r_0 has been called the "Sisyphus" distance, apparently in some obscure allusion to the corresponding ancient Greek myth. Despite its elegance the model cannot account satisfactorily for many of the causes of capture. Particularly for the case of mobile users, the causes of capture cannot be accounted for in a deterministic way. Thus let us propose a simple probabilistic model and explore its ramifications.

Let

g^{n-1}= probability that one packet is successfully received when there are n packets simultaneously transmitted ($0\leq g\leq 1$, $n=1,2,\ldots$). Obviously this is a simple model but it is a useful way to begin. Let us examine the behavior of the standard slotted ALOHA scheme under our model. Let S be the probability that a packet gets successfully transmitted in a given slot. This probability is given by

$$Ge^{-G} + \frac{G}{2!} e^{-G}g + \ldots + \frac{G^k}{k!} e^{-G} g^{k-1} + \ldots$$

where G is the rate of the channel traffic process (assumed to be Poisson) and the slot length is assumed to be equal to the packet length (taken to be equal to one).

Thus S, which also represents the average throughput rate, is given by

$$S = \frac{e^{-G}}{g} \left[Gg + \frac{(Gg)^2}{2!} + \ldots\right] = \frac{e^{-G}}{g} \left[e^{Gg} - 1\right] = \frac{e^{-G(1-g)} - e^{-G}}{g} .$$

This relationship implies a very sensible behavior as depicted in Fig. 5. For $g=0$ we obtain $S=Ge^{-G}$ which is as expected. For $g=1$ we obtain $S=1-e^{-G}$ which is also expected for perfect capture. For $0<g<1$ we obtain

$$G_{max} = \frac{1}{g} \ln \frac{1}{1-g}$$

which is achieved for $S_{max} = (1-g)^{\frac{1-g}{g}}$ and which approaches 1 for $g \to 1$, and e^{-1} for $g \to 0$, as expected.

Now we would like to turn our attention to the case of random access with frequency hopping and study the effect of capture on it. A useful specific model for a frequency-hopped random access scheme recently proposed by B. Hajek [9] is the following. Each packet consists of n bytes and each byte is transmitted in one of q frequency bins chosen randomly and independently by the corresponding user. Each packet is a

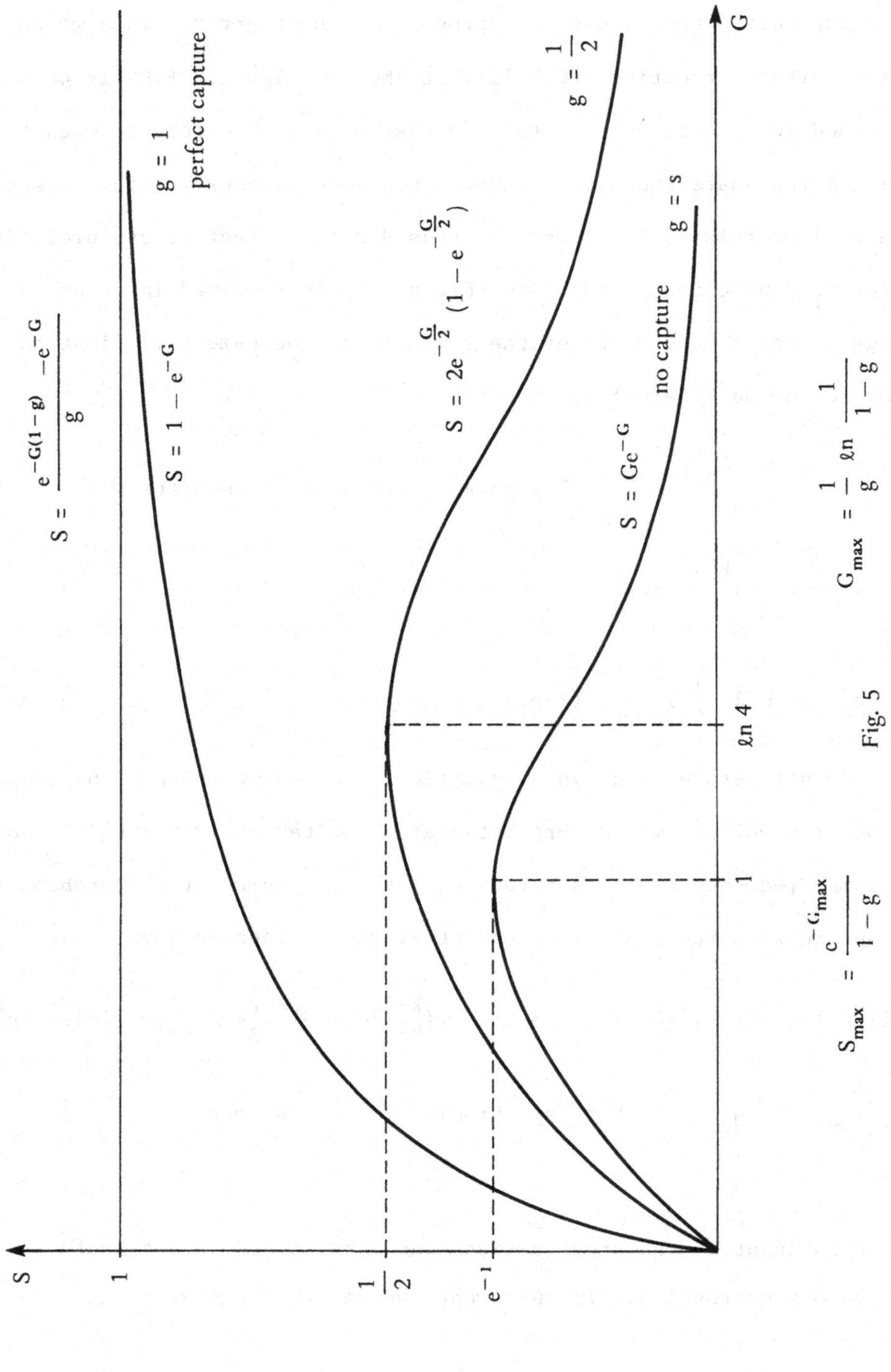

S
1
1/2
e^{-1}
G
1
ℓn 4
$S = \frac{e^{-G(1-g)} - e^{-G}}{g}$
g = 1
perfect capture
$S = 1 - e^{-G}$
$S = 2e^{-\frac{G}{2}}(1 - e^{-\frac{G}{2}})$
$g = \frac{1}{2}$
$S = Ge^{-G}$
no capture
g = s
$G_{max} = \frac{1}{g}\,\ell n\,\frac{1}{1-g}$
$S_{max} = \frac{e^{-G_{max}}}{1-g}$

Fig. 5

word from a Reed-Solomon code such that if up to t bytes are received in error the entire packet can be reproduced without error. Thus we build-in some error correction capability at the byte-level. Finally we make the standard assumption of total and destructive interference when two or more bytes share the same frequency bin simultaneously. This assumption will be relaxed later when we consider the effect of capture. If we let p_1 denote the probability that a byte is received in error even though no other byte occupies the same bin at the same time (just to allow for noise effects) we obtain

$$p_k = Pr[\text{byte success}/_{k} \text{ packets are being transmitted}] =$$

$$= \left(1 - \frac{1}{q}\right)^{k-1} (1-p_1)$$

or

$$p_k = \left[\left(1-\frac{1}{q}\right)^2\right]^{k-1} (1-p_1)$$

if we do not assume byte synchronization. Let us now assume that capture may occur whenever two or more bytes are simultaneously transmitted at the same frequency bin. As before we let g^{i-1} represent the probability of capture when there are i bytes interfering. Then we have

$$p_k=(1-p_1)\left[Q^{k-1}+\binom{k-1}{1}Q^{k-2}(1-Q)\tfrac{1}{2}\,g+\ldots+\binom{k-1}{k-2}Q(1-Q)^{k-2}g^{k-2}\frac{1}{k-1}+(1-Q)^{k-1}g^{k-1}\frac{1}{k}\right]$$

$$\text{where } Q=\begin{cases} 1-\dfrac{1}{q}, & \text{if byte synchronization is assumed} \\ \left(1-\dfrac{1}{q}\right)^2, & \text{else} \end{cases}$$

Unfortunately the above series cannot be summed into a useful closed-form expression. In any event, we can then calculate

$$P(k) = [\Pr\ \underline{\text{packet}}\ \text{success}/_{k\ \text{packets being transmitted}}]$$

$$= \Pr\ [\text{at most } t \text{ byte errors}/_{k\ \text{packets}}]$$

$$= \sum_{i=0}^{t} p_k^{\,n-i}(1-p_k)^i \binom{n}{i}$$

which for t=0 (no coding) reduces to $p_k^{\,n}$ as it should. Finally we obtain that

$$\Pr\ [\text{packet success}] = \sum_{k=1}^{\infty} e^{-G}\,\frac{G^k}{k!}\,P(k)$$

assuming a Poisson channel traffic process, with

G = average number of packets transmitted per time slot

$= \lambda q$

where λ represents the channel traffic rate <u>per frequency bin</u>.

Another way of computing throughput, suited to the case of n=1, and t=0, is the following. Let

$$p_k' = \Pr\ [\text{success of a byte in a } \underline{\text{given}} \text{ frequency bin}/_{k\ \text{packets}}] =$$

$$= \binom{k}{1}(1-Q)Q^{k-1} + \binom{k}{2}(1-Q)^2Q^{k-2}g + \ldots + \binom{k}{k-1}(1-Q)^{k-1}Qg^{k-2} +$$

$$+ (1-Q)^k g^{k-1} =$$

$$= \frac{1}{g}\left[\binom{k}{1}(1-Q)Q^{k-1}g + \ldots + (1-Q)^k g^k\right] =$$

$$= \frac{((1-Q)g+Q)^k - Q^k}{g}$$

Then the throughput <u>per bin</u> is given by

$$S' = \frac{S}{q} = \sum_{k=0}^{\infty} \frac{G^k}{k!} e^{-G} \frac{((1-0)g+0)^k - 0^k}{g}$$

$$= \frac{e^{-(1-g)\frac{G}{q}} - e^{-\frac{G}{q}}}{g}$$

If q=1 we obtain the previously derived Aloha result. Of course for any value of q the relationship between throughput per bin and channel traffic per bin is the same as in the case of Aloha. This is expected since when n=1 and t=0 the random and independent choice of bins for each byte (packet) implies the parallel, independent operation of q Aloha channels.

Of course the discussion of capture and its effects in frequency hopping and signalling schemes is far from complete. Our model (eventhough an improvement over earlier deterministic ones) is too simple. Also no mention is made of retransmission schemes and of availability of realistic feedback information on the fate of transmitted packets. In fact the topic of feedback will be the subject of the next lecture. It is hopefully clear, however, that capture offers an intriguing possibility of exploring its use in securing radio transmissions in the presence of interference (friendly or hostile).

Lecture 3: Acknowledgements and Feedback Information

Another basic assumption in the study of most multiple access protocols is the existence of perfect ternary feedback. This is to say that every user is able to determine with some delay (which is often assumed to be negligible) whether during a slot there were zero, one, or more than one packets transmitted (this is accomplished by monitoring the channel) and, furthermore, he is able to determine whether his transmissions are successfully received (either via this ternary feedback when noise is negligible or via acknowledgement message mechanisms).

Acknowledgement messages are generally important because they can be used for flow control in addition to their error control function implied so far. Also they are often indispensable in order for the useful ternary feedback assumption to be valid; that is, monitoring of the channel by the transmitting node is in practice insufficient to determine whether successful reception took place.

In several applications acknowledgement messages are not permissible or are not reliable. For example some nodes are passive receivers in order to conceal their location. Other nodes may be very near interfering sources and thus can serve as "transmit-only" nodes. In such cases it is important to devise other schemes for securing the delivery of radio messages. Conceptually it is easiest to start with the assumption that there can be no reliable feedback at all. That is, there are no acknowledgements and there is no channel monitoring. This is the case we want to examine here.

In [13a] an information theoretic approach to the problem establishes the feasibility of achieving the maximum possible throughput rate permitted by the basic multiple access protocol used provided there is enough redundancy at the packet level and sufficient tolerance of delay. We do not simply mean use of forward error correction techniques but rather randomized retransmissions of packets or, more generally, transmissions of redundant packets.

Here we focus on specific schemes that employ the above philosophy and are easy to implement, and we seek to establish their performance. We consider three policies. The first one assumes random access, with each packet being retransmitted randomly, again and again, until a fixed period of time W from the time of its initial generation elapses. The second policy is identical to the first one except that the retransmissions continue randomly until a fixed number M of them is completed. Assuming the rescheduling times are independent and exponentially distributed with parameter λ and setting $\lambda W=M$ we can consider an obviously superior third policy that requires the retransmissions to continue either until M are completed or until W seconds pass, whichever occurs first. We want to calculate three quantities for each plicy: 1) The channel traffic induced by the policy, 2) the throughput achieved, and 3) the probability of successful delivery of a packet. Let us assume that there are K users, each generating <u>new</u> packets according to a Poisson process of rate λ. The retransmissionscontinue independently using the same parameter λ.

1) <u>Channel Traffic</u>

a) <u>First Policy</u>: Let us look at ome of the K users. Let N_t denote the number of new packets generated by him in the time interval $(0,t]$ and let g_t denote the total number of packet transmissions or retransmissions in $(0,t]$. Then

$$E\,[dg_t/\text{past until } t] = P[dg_t = 1/\text{past until } t] =$$

$$= \lambda dt + (N_t - N_{t-W})\lambda\, dt$$

$$E\,[dg_t] = \lambda dt + \lambda^2 W dt$$

$$g = \frac{E[dg_t]}{dt} = \lambda + \lambda^2 W$$

and

$$G = Kg = K\lambda(1+\lambda W)$$

while the input is only $K\lambda$.

b) Second Policy: Here we need the notion of the number of "active" packets at time t. Let us denote it by N'_t. In the first policy N'_t was simply $N_t - N_{t-W}$. Here it is

$$N'_t = N_t - k$$

where k is defined by

$$\sum_{i=1}^{k} \tau_i < t - \frac{M}{\lambda} \leq \sum_{i=1}^{k+1} \tau_i$$

where τ_i is the ith interarrival time between new packets of the user. Thus

$$E\,[dg_t/\text{past of } N_t \text{ and of } g_t] = P\,[dg_t=1/\text{pasts of } N_t \text{ and } g_t] =$$

$$= \lambda dt + N'_t \lambda dt = \lambda dt + (N_t - k)\,\lambda dt$$

Then

$$E\,[dg_t/\text{past of } N_t \text{ only}] = N_t \frac{t - \frac{M}{\lambda}}{t}$$ (this is due to the fact that for given N_t the arrival points in $(0,t]$ are uniformly distributed [13b]). Finally

$$E\,[dg_t] = (\lambda + M\lambda)dt$$

and

$g = \lambda(1+M)$ or $G = K\lambda(1+M)$

c) Third Policy: Choose W and M such that $M = \lambda W$.

Let N'_t = number of packets "born" within the last W seconds

N''_t = number of new packets born within the W window that have not completed M retransmissions.

Again

$$E\,[dg_t/\text{past}] = \lambda dt + N''_t\,\lambda dt$$

and

$$E\,[dg_t/\text{past of } N_t \text{ only}] = \lambda dt + E[N''_t/\text{Past of } N_t]\lambda dt$$

$$= \lambda dt + (N'_t - k)\,\lambda dt$$

where k = number of packets for which $t - t_i > E[X]$ and where t_i is the arrival instant of the (i^{th}) packet and X is the "time-to-death" of an active packet (a r.v. equal to the sum of M i.i.d. exponential variables truncated at the value W). We find that

$$E\,[X] = Wq_W - \frac{M}{\lambda}\frac{(\lambda W)^M}{M!}e^{-\lambda W} + \frac{M}{\lambda}(1-q_W)$$

where

$$q_W = \int_W^\infty e^{-\lambda w}\frac{(\lambda x)^{M-1}}{(M-1)!}\lambda dx = P\;[\text{less than M attempts in W seconds}]$$

$$= e^{-\lambda W}[1+\lambda W + \frac{(\lambda W)^2}{2} + \dots + \frac{(\lambda W)^{M-1}}{(M-1)!}]$$

Thus

$$E\,[dg_t/N_t] = \lambda dt + N_t\frac{E[x]}{t}\lambda dt$$

and

$$E\,[dg_t] = \lambda dt + \lambda^2 E[x]dt$$

Combining the above equations we finally obtain

$$G = K\lambda\ (1+M(1- \frac{M^M}{M!}\ e^{-M}))$$

2) <u>Throughput</u>

We need an approximation due to the analytical complexity of the precise expressions for point processes with time-varying rates. Namely we assume the packet duration to be small relative to $\frac{1}{\lambda}$ (a perfectly reasonable assumption for a large number of users and in order for the self-interference due to overlapping transmissions by the same user to be small). We take ε to be one. Thus λ must be very small (besides $K\lambda$ must be <1 anyway).

a) <u>First Policy</u>. Given the past history of all transmissions the probability of a successful transmission (or retransmission) by one user at time t is given by

$(\lambda + \lambda(N_t - N_{t-W}))dt$. P [no other user initiates a transmission within (t-1, t+1]].

The transmission processes of all users are independent but the probability of zero transmissions in an interval for a process with time-varying rate is given by

$$e^{-\int_{t-1}^{t+1} \lambda_\tau d\tau} = e^{-\lambda\int_{t-1}^{t+1} (N_\tau - N_{\tau-W}) d\tau}$$

Because of our assumption, $N_\tau - N_{\tau-W}$ is approximately constant for $\tau\varepsilon(t-1, t+1]$. Thus, if by s we denote the throughput for one user, we have

$$sdt = \lambda(1+\lambda W)dt\ [e^{-2\lambda}E(e^{-\lambda 2(N_t - N_{t-W})})]^{K-1}$$

or

$$s = \lambda(1+\lambda W)\ [e^{-2\lambda} e^{-\lambda W(1-e^{-2\lambda})}]^{K-1}$$

and for small λ

$$1-e^{-2\lambda} \approx 2\lambda$$

thus

$$s = \underbrace{\lambda(1+\lambda W)}_{g} \quad e^{-2\overbrace{\lambda(1+\lambda W)}^{g}(K-1)}$$

For slotted random access we can get rid of the 2 factor throughout.

Let

S = s.K and G = s.K. Thus

$$S = Ge^{-G\frac{K-1}{K}}$$

which yields

$$S_{max} = \frac{K}{K-1}\ e^{-1} \text{ for } G_{max} = \frac{K}{K-1}$$

and optimum window size

$$W_0 = \frac{\frac{1}{(K-1)\lambda} - 1}{\lambda}$$

for any $\lambda < \frac{1}{K-1}$. Since a successful transmission may be one of a previously successfully transmitted packet we must adjust the value of throughput to its "real" level by dividing by $1+\lambda W$.

Thus

$$S_{real} = K\lambda e^{-(K-1)\lambda(1+\lambda W)}$$

which is always (for any λ) inferior to the level achieved by plain random access (W=0 and $S=K\lambda e^{-(K-1)\lambda}$ as opposed to $K\lambda e^{-1}$). The explanation of the apparent paradox is that the "real" throughput is measured as rate of "true" successes in a given time period, while this policy involves possible successes later on, after, sometimes considerable, retransmission delay. We shall see that the probability of success of a given packet increases with our policy. Thus we will confirm the conceptual result from information theory that guarantees successful deliveries up to a maximum rate but with increasing delay.

b) Second Policy. The approach is exactly as before. Because of our approximation we can reduce the result to

$$sdt = \lambda(1+M)dt\ e^{-\lambda(1+M)(K-1)}$$

and

$$S = Ge^{-G\frac{K-1}{K}}$$

with

$$S_{max} = \frac{K}{K-1}\ e^{-1} \text{ for } G_{max} = \frac{K}{K-1} \text{ and } M_0 = \frac{1}{(K-1)\lambda} -1$$

Again, the determination of the real throughphput requires an adjustment and yields

$$S_{real} = K\lambda\ e^{-\lambda(1+M)(K-1)}$$

which is again worse than plain random access without retransmissions. Of course the same explanation applies here as before.

c) Third Policy. For $\lambda W = M$ the expressions for S and S_{real} are exactly as before except that G is given by the appropriate expression corresponding to the third policy.

3) Success Probability

Obviously retransmission methods must enhance the chances of the successful transmission of a given packet if overloading of the channel is avoided. We can confirm this expectation by analyzing the proposed policies.

a) First Policy. First note that with plain random access (W=0) the success probability is $e^{-\lambda(K-1)}$. For W>0 we calculate it by conditioning on the number x of retransmissions afforded within the fixed window W. Obviously x is a Poisson random variable.

$$P\,[\text{success}] = 1-P\,[\text{all collisions}] = 1 - \sum_{x=0}^{\infty} P\,[x \text{ retransmissions}].$$

$$.P\left[\begin{array}{l}\text{collision in all x and}\\ \text{in original transmission}\end{array}\right] = 1 - \sum_{x=0}^{\infty} \frac{(\lambda W)^x}{x!} e^{-\lambda W} \cdot \rho^{x+1}$$

where

$$\rho = \Pr\,[\text{collision in single try}] = 1-e^{-\lambda(1+\lambda W)(K-1)}$$

After performing the summation we find

$$P\,[\text{success}] = 1 - \rho e^{-\lambda W(1-\rho)}$$

We can easily determine that this expression yields a better value (for $\lambda < \frac{1}{K-1}$) than the one for W = 0.

b) Second Policy. A similar calculation as before yields

$$P\,[\text{success}] = 1 - (1-e^{-\lambda(1+M)(K-1)})^{M+1}$$

which, again, is greater than $e^{-\lambda(K-1)}$

c) Third Policy. The actual calculation here is much more complex than before. However if $M=\lambda W$ we can compare the first and the second policies. This amounts to comparing

$$(1 - e^{-\lambda(K-1)(M+1)})^M \quad \text{vs} \quad e^{-Me^{-\lambda(K-1)(M+1)}}$$

or

$$(1-x^M)^M \quad \text{vs} \quad e^{-Mx^M}$$

for $x = e^{-(K-1)}$. If λ is very small (as assumed), x is close to one and then we can verify that the quantity on the left is less than the one on the right and thus the second policy is preferable. Of course detailed study of these curves can establish what happens for higher values of λ and whether there is a crossing of the two curves as M increases.

Of course it is of interest to explore other retransmission policies as well and to incorporate in their study the effect of some form of feedback as well as that of the capture phenomenon discussed in the previous lecture.

Lecture 4: A Case Study

The totality, complexity, and interaction of security issues along with other operational issues in the design of a mobile-user (generally multi-hop) radio network makes the task of designing such a system a formidable one. Without the benefit of sufficient theoretical support for globally modeling such a design problem we are often forced to make hard choices based on qualitative and uncertain guidelines. In this lecture we would like to explore one example of such a design. The example is general enough to be useful and specific enough to put to a test the capabilities of the different theories and approaches that have been proposed in connection with mobile radio network design.

Suppose that we are asked to specify the rules of transmission and retransmission, choice of waveform, rules of error and flow control, relaying, etc. for a finite number of users operating under power and equipment constraints and moving over a large geographical area without central control or guidance and in the presence of other hostile users.

the first question to ask is what performance criteria or measures are to be used. The answer shows how complex the problem is right from the beginning. There are several performance measures. We may classify them in the following way.

1) Measures of Effectiveness

a) average delay per message; this is defined as the average time from generation of a message to reception by the intended receiver. Obviously this time includes several "hop" cycles of queueing, processing, propagation, and transmission times if there is relaying in the network. Theoretical evaluation of delay is very difficult in realistic environments due to the highly complex nature of interacting queueing systems.

b) throughput; this is related to the delay but in a complicated way. It is defined as the average total rate of successfully delivered messages in the network. It is not clear that there are protocols that minimize delay and at the same time maximize throughput as we would like them to do.

c) stability; this is even difficult to define. Contention based protocols, such as ALOHA, sometimes display a dynamic behavior that results in total degradation of performance, namely a reduction of the throughput to the lowest levels and a simultaneous increase of the delay to intolerable levels. Such behavior is called unstable and is obviously unacceptable.

2) Measures of Efficiency

This relates in a vague sense to the essence of the entire field of networking, namely the shared use of resources. How efficiently is the bandwidth utilized? How often are there wasted time or frequency slots due to collisions or due to idleness of users to whom they are dedicated? This is a subtle concept not directly related to the preceding criteria.

3) Measures of Survivability

a) robustness; this requires the maintenance of satisfactory performance over a wide range of values of the critical parameters of the network such as traffic statistics, number of users, topological layout, etc. Often robustness cannot be achieved without adaptability.

b) adaptability; when the critical parameters of the network are time varying it is desirable or, sometimes, imperative to adapt the protocol to the changes in order to maintain satisfactory performance. Robustness and adaptability together constitute ingredients of survivability and graceful degradation. The latter also requires fail-safety.

c) fail-safety; when nodes or links fail due to jamming or whatever other reason, the ability of the network to operate must not be impaired. This is perhaps the most fundamental requirement for certain networks. It is also an overriding criterion, since it may be incompatible with minimizing delay, maximizing throughput, and maintaining high efficiency. It requires that the protocol rules are such that no deadlocks due to errors or data base inconsistencies occur and that no node or link is indispensable. The latter requirement often dictates a distributed control architecture (ruling out, for example, polling schemes) and it implies the former, since, under distributed control operation, it is possible that inconsistencies, instabilities, and deadlocks may occur.

The awesome complexity of the total problem suggests a blend of rigor and "wisdom" in the approach or a mixture of science and "art" or, plainly, an engineering solution. Thus let us start with an envisioned architecture as in Fig. 6. We would like to invent a distributed algorithm that will allow the nodes to connect themselves as depicted in the figure. We consider two such procedures (very similar to each other). We describe them first in a fictitious centralized version and then in the implementable, distributed one.

1st Method (Centralized Version)

This method produces the node clusters shown in Fig. 7. The nodes are first numbered from 1 to N. The central controller starts with the highest numbered node, say node N, and declares it a cluster head. Then it draws a circle around that node N with radius equal to the range of communication. The nodes inside the circle form the first cluster. It then considers whether there are nodes outside this circle. If there are, it tentatively considers drawing a circle about node N-1. Should any nodes lie within this circle that were not already within the first circle, node N-1 becomes a cluster head and a circle is drawn about it. Then consideration of tentative cluster head status for nodes N-2, N-3, etc. follows, until all nodes lie within at least one circle. The resulting arrangement provides every node with a cluster head. Any pair of clusters may be directly linked, they may even cover one another, they may simply overlap, or they may be disconnected. In the last two cases, selected nodes must serve as gateways for the interconnection of the cluster heads. This issue will be addressed in the discussion of the distributed version of the algorithm.

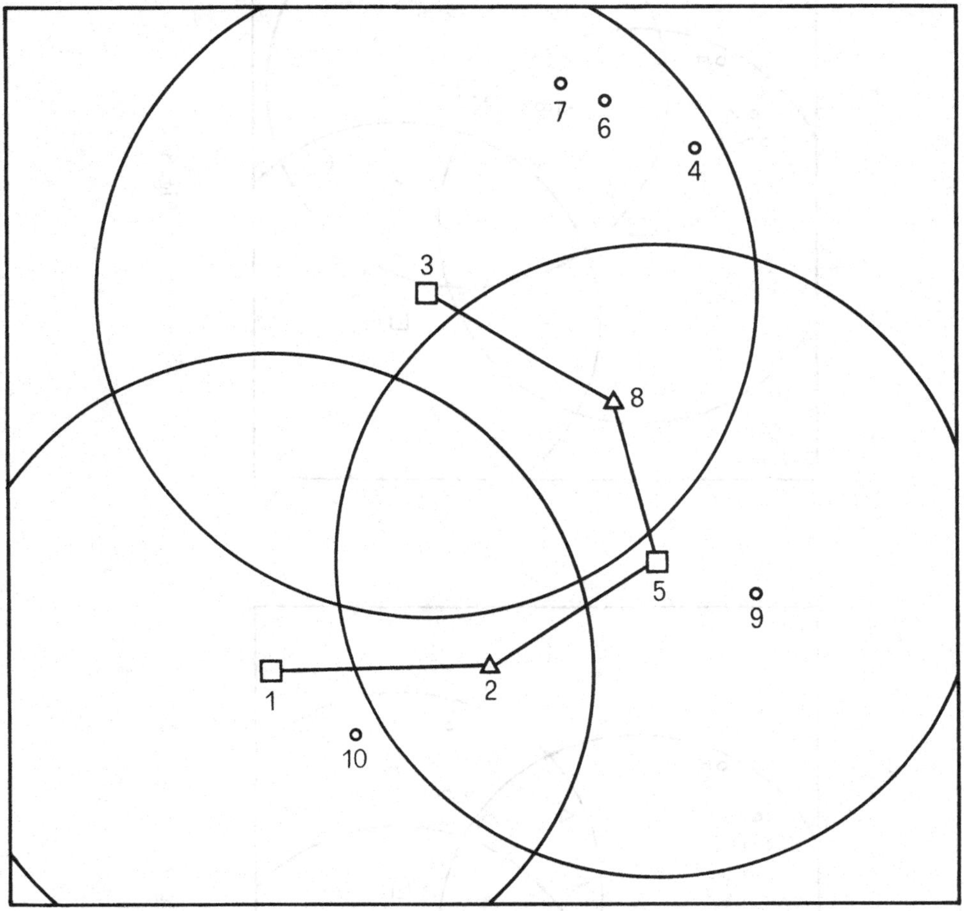

Fig. 6

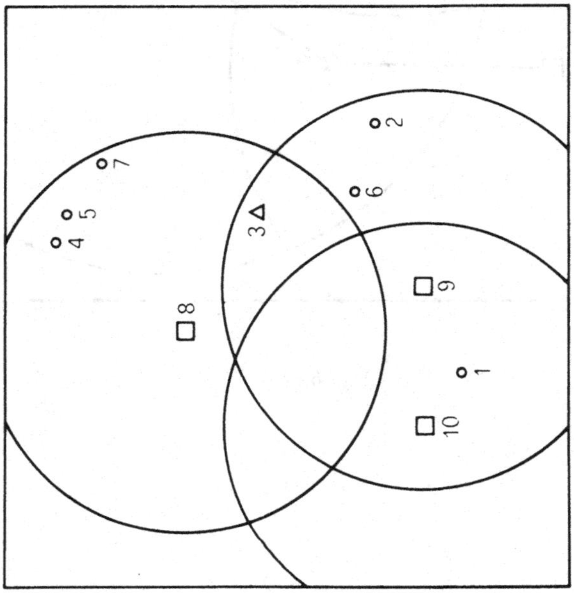

Fig. 7

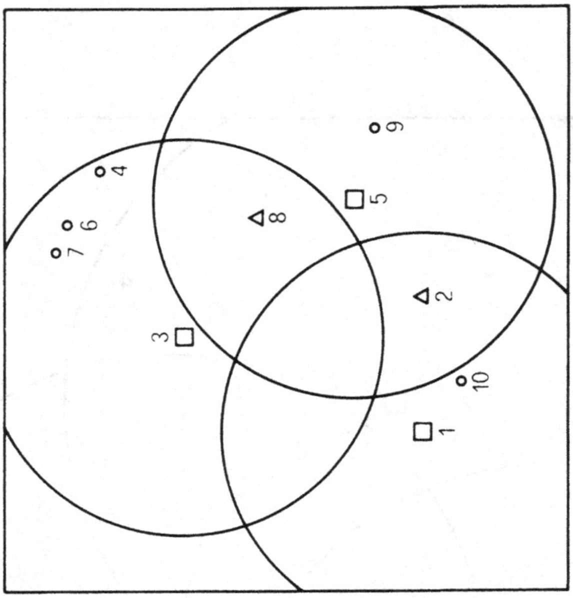

Fig. 8

2nd Method (Centralized Version)

In the alternative method the procedure is a slight variation of the one just described. The central controller starts with the lowest numbered node, node 1, declares it a cluster head, and draws a circle around it with radius equal to the fixed communication range, thus forming the first cluster. If node 2 lies in this circle it does not become a cluster head. If not, it does become a head and the controller draws a circle around it. Continuing in this manner; node i becomes a cluster head unless it lies in one of the circles drawn around earlier nodes. Unlike the previously described case, here no cluster can cover another nor can two clusters be directly linked. To facilitate comparisons the nodes are numbered in reverse order from that shown for the 1st method. Thus, nodes 1,2,3, etc. of the 1st method are nodes N,N-1,N-2, etc. of the corresponding 2nd method in Fig. 8.

We now proceed to describe the distributed versions of these algorithms, the details of which can be found in [14] .

1st Method (Distributed Version)

The algorithm has two logical stages: first, the formation of clusters and second, the linking of the clusters. Each node performs the steps of the algorithm based on local information. Thus some simple message exchange is necessary. If there are N nodes we consider two TDMA frames each consisting of N slots. In frame 1, in its assigned ith slot, node i broadcasts the identities of the nodes it has heard from during the earlier slots of this frame. (Thus node i also receives partial connectivity information of the nodes that it can hear.) So by the end of this frame, node i has filled in some of the entries of its connectivity matrix. In particular, it can fill in the elements above the main diagonal, i.e., the elements (i,j) of the ith row that satisfy j>i. The element (i,j) is set equal to 1 if:

a) node i heard from node j

and

b) node i appears in the connectivity list broadcast by node j.

In frame 2, each node broadcasts in its assigned slot its full connectivity row. This is possible because node i has completely filled in the ith row of the connectivity matrix by the time of the ith slot of Frame 2. Here is how node i determines the bidirectionality of links (i,j) for j<i. Node i sets connectivity matrix element (i,j), for j<i, equal to 1, if the ith element of the connectivity row received from node j during Frame 2 is equal to 1. By the end of the frame each node knows the two-way connectivities for itself and for its neighbors. The global connectivity matrix is not available to every individual node - only a partial version of it is formed by each node. However, for the case of error-free transmissions, all versions are consistent with the global true matrix.

Now the clusters can be formed. At the ith slot of the second frame, node i can perform a logical function that permits it to determine whether it is a head or an ordinary node, which it can then transmit along with its row connectivity. We use the rule that the node with the highest identity number among a group of nodes is the first candidate to claim cluster head status. Thus node i first checks its own connectivity row. If there is no neighbor with higher identity number, node i becomes a cluster head. If another neighbor exists with higher identity number, that neighbor will become a cluster head, so i doesn't have to. However i must also check whether it is the "highest" neighbor of some other node $j < i$. This can be done by checking the received connectivity rows from the lower numbered neighbors. If node i is the highest in some row $j < i$, node i must become a cluster head for at least node j. Thus node i is able to broadcast in the ith slot of the 2nd frame his status. This information is needed for the linking of the clusters as will be seen later.

At the end of the second frame each node knows all head nodes that are one hop away and some, but perhaps not all, heads that are two hops away. Thus, by the end of the second frame, clusters are formed and the data base necessary for the second logical function of the algorithm (the linking of the clusters) is available. The linking is accomplished by the introduction of Gateway nodes. Every non-head node is a candidate to become a gateway. There are three cases to be considered. The first case is shown in Figure 9a. Here there is no need for gateways since the heads of the clusters are directly linked. The second case is depicted in Figure 9b. Here exactly one node is needed to link up the two heads. Clearly the candidates are the nodes in the intersection of the two cluster regions. The third case is pictured in Figure 9c. Here at least two nodes (one from each cluster) are needed. It is, of course, assumed that suitable such nodes exist; otherwise the net cannot be connected. In the sequel we describe the procedures used to achieve the link-up in the last two cases. Every node, which is not a cluster head, is a candidate gateway node. Each pair of heads in a node's list of heads that are one hop away corresponds to a pair of overlapping

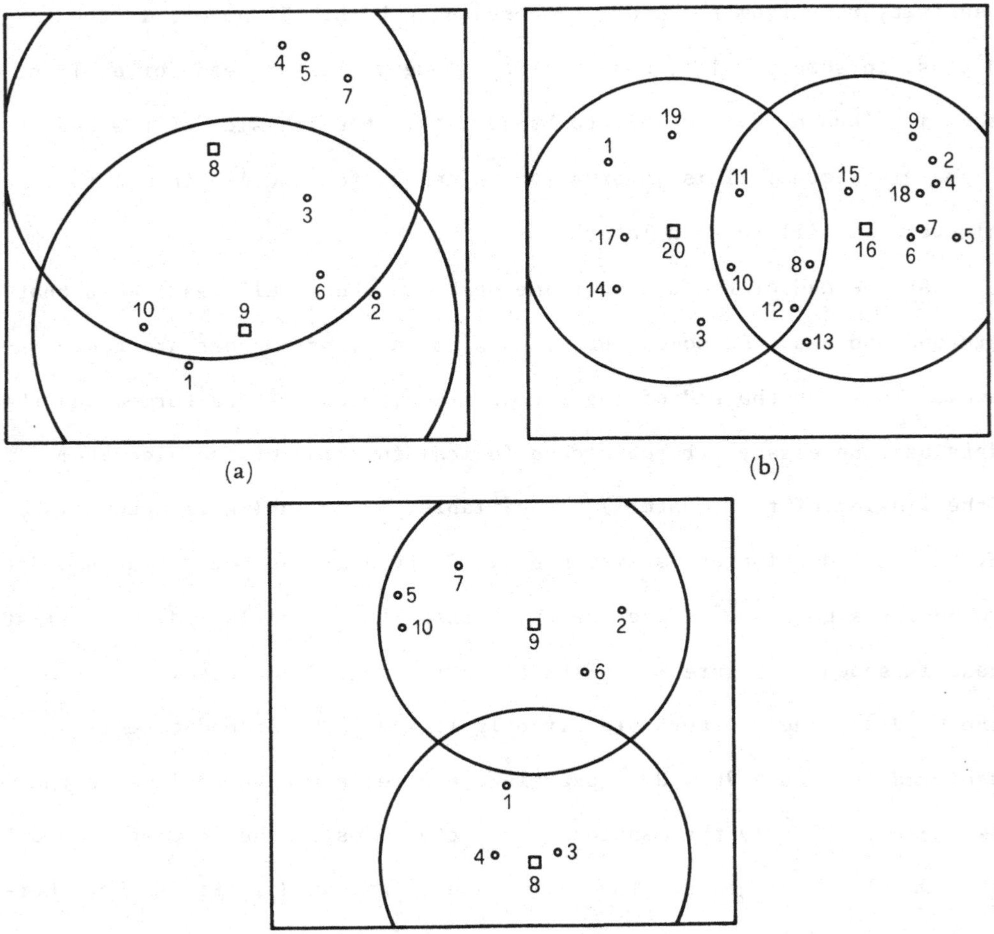
(a)
1 2 3 4 5 6 7 8 9 10
(b)
1 2 3 4 5 6 7 8 9 10 11 12 13 14 15 16 17 18 19 20
(c)
1 2 3 4 5 6 7 8 9 10

Fig. 9

clusters are already linked through another cluster head. If an unlinked pair of heads is found, the examining node is a candidate gateway for linking these heads. The highest numbered node in the intersection of the two clusters is chosen to become a gateway for that pair. All nodes in this intersection are aware of each other since they can be at most two hops away from each other and every node possesses the connectivity row information for every one of its neighbors. Thus there is no ambiguity in the selection of the gateway node.

Case of non-overlapping clusters: Cluster head pairs consisting of a node's own head plus a head from the node's list of heads that are two hops away identify non-overlapping clusters. For linking up two clusters that do not overlap, at least one node from each cluster must become a gateway. Each node proceeds to examine every possible pair of nodes, the first member of which is its own cluster head and the second member of which is a cluster head in its list of heads that are two hops away. To avoid creation of redundant gateways the node attempts to ascertain the need for the creation of a gateway by checking, for each such pair, whether a path may have been created through another cluster head. Thus it seeks nodes among its neighbors that may include in their connectivity rows the second member of the headpair it is examining, together with a cluster head from its list of nodes one hop away. If no such circumstance is established however, the node proceeds further, assuming that its services are needed for this linkage. There may be several pairs of potential gateway nodes that can link two clusters. Each node may be aware of several of those, but perhaps not all of them. The (arbitrary) deterministic rule chosen for resolving the ambiguity is to select the pair with the largest sum of identity numbers. In case of a tie the pair involving the node with the highest number is chosen. Unlike the previous case here we may end up with extra gateways with two or perhaps more pairs becoming gateways. It is worth noting however that such multiple linkage outcomes are not very likely for most topological configurations. In some cases only one of the two potential gateway nodes in a pair may decide to become a gateway while the other may find that it is not needed if another pair of higher numbered nodes is available and known to it. The existence of such a pair may not be known to it. The existence of such a pair may not be known to both partners of the first pair, and thus asymmetric situations can arise. Such outcomes, however, are rare and only a harmless nuisance. They need not affect the network's operation and cannot be avoided without substantially increasing the data bases available at each node by additional messages exchanges.

2nd Method (Distributed Version)

The two methods have nearly identical implementations. Both use the same data structures, and both follow the same transmission schedule described earlier. The only difference is that the rules for forming the clusters and for assigning cluster heads are different. The difference has already been described in the centralized version. In the distributed implementation, the rule is simply that a node becomes a cluster head unless it has a lower numbered cluster head for a neighbor Thus instead of announcing its node status, each node broadcasts the identity of its own cluster head during Frame 2 transmission. This enables nodes to fill in both their lists of heads one and two hops away.

Once the clustering has been formed the "potential" links between neighbors must be activated in a coordinated but distributed way.

Essentially, each node attempts to make an arbitrary assignment of slots to its neighbors and thus form a TDMA schedule. Obviously, each neighbor will attempt to do likewise, but as it may have a different number of neighbors than the original node and as these neighbors set up their own schedules also, inconsistencies will in general arise that must be resolved in order to arrive at consistent, conflict-free assignments. A systematic method in which these schedules are set up and the conflicts resolved is needed.

The method consists of two activities, namely, allocating slots and resolving scheduling conflicts. Slot allocating occurs during frame 1 and conflict resolving occurs in frame 2. These frames are the ones corresponding to the previous algorithm. Broadcasts of information relevant to the present method take place during the same slots that information relating to the previous algorithm is sent.

The first slot of each schedule is arbitrarily selected as a broadcast slot. During this slot each node monitors the transmission from its own clusterhead. It is during this slot that any node can access any neighboring clusterhead using random access techniques. Other time slots are allocated for the activation of specific links. If the link between a pair of nodes is bidirectional, the highest numbered of the two nodes is responsible for allocating the slot. If a link is unidirectional, the receiving node must allocate the slot. When a node allocates a slot, it chooses the earliest slot available for that link. At the end of frame 1 every link has been allocated one slot.

During frame 1 each node broadcasts, in turn, its current link activation schedule, which will contain only those nodes heard from, thus far, during frame 1. As other nodes receive the broadcasts of these schedules, they either allocate a slot, if this is a unidirectional link, or simply update their own schedule, if this link is bidirectional. An update involves a possible change in one's own schedule to make it consistent with information just received.

At the end of frame 1, each node has its own version of the link activation schedule. During frame 2 the conflicts that may exist can be resolved following a simple procedure described in [15].

Thus we have some means of connecting nodes that were previously disconnected and of providing a backbone structure for their communication. The method is distributed and thus relatively secure. Many issues remain of course to be addressed in order to fully describe the operation of the network. The most important step however has been taken. In the next lecture we will show how such a network is inherently robust against jamming threats.

Lecture 5: Inherent Network Security

It has been known [16,17] that one way of providing protection to a single communication link is via the introduction of relays. This protection is twofold. One, resistance to interference is achieved for fixed transmitter power, and two, resistance to interception is achieved by lowering the transmission power. Thus if

R_J = distance from interfering source to receiver

R_L = effective transmission distance from transmitter

we have

$$R_L = R_J K^{\frac{1}{2}}$$

where K is a composite coefficient that depends on processing gain, bandwidth, bit duration, power, antenna configuration, and system losses. The precise form of the relation between R_J and R_L may vary for more complex channels. If K or R_J cannot be changed, we can still increase R_L by introduction of relays. By a relay we mean an intermediate node that decodes, re-encodes, and transmits as opposed to one that simply "repeats" the received electromagnetic waveform.

Suppose that originally we have the setup of Fig.10 and then that of Fig. 11 corresponding to a canonic in-line configuration of relays.

$$R_d = S - R_J = R_J \ [(1+K^{\frac{1}{2}})^{n+1} - 1\ ]$$

Thus

$$n \leq \frac{\log\ [1 + R_d/R_J]}{\log\ (1 + K^{\frac{1}{2}})} - 1$$

for a desired R_d/R_J. More intricate analysis is possible for the optimum location of a given number of relays in order to maximize R_d/R_J. A similar approach can be applied to the study of the anti-intercept function of relaying.

Our purpose here is to show that a network has an inherent capability of providing strong resistance to threats dur to its natural structure that includes multiple relays. In particular, we shall use the example of the network considered in the last lecture to demonstrate this capability.

We consider an arbitrary, but not inherently disconnected, set of

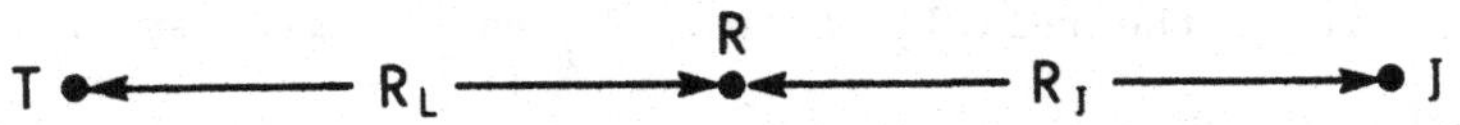

Fig. 10

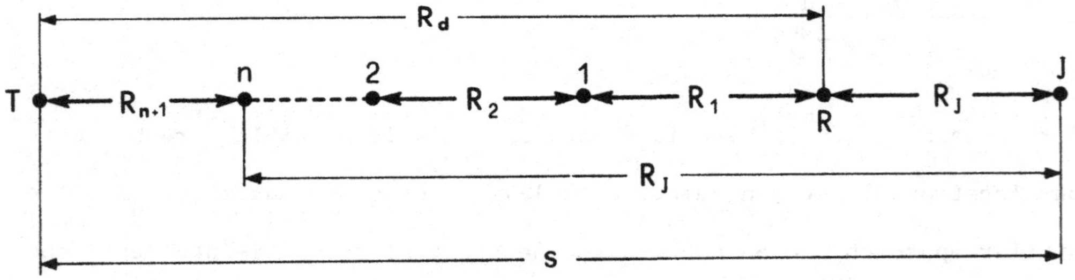

Fig. 11

N radio nodes that move about and use the algorithm described in the last lecture to organize themselves into overlapping clusters.

These network structuring algorithms enable the nodes to self-organize into a network structure that provides anti-jam protection through the use of relaying. The selection or relays is done automatically and without the need for a central controller. In addition, the network structure is continually self-adapting in response to a changing jamming environment.

To illustrate this self-organizing capability, we used a digital computer simulation model. In the simulator, node connectivities are computed based on some fixed propagation model [18].

Our example network is the one shown in Figure 6. The bidirectional connectivities for this network are shown in Figure 12. The radio connectivities shown are for the case when no jammers are present.

In the absence of jamming, the network self-organizes into the set of node clusters shown in Figure 6; three clusters are formed with heads at nodes 1, 3, and 5. The circles indicate the communication range for each head, and every node is located within one of these clusters. Nodes 2 and 8 have become gateways (relays) to join together the cluster heads, forming the "backbone" network.

The sequence of frames in Figure 13 illustrates how the network re-structures itself in response to a changing jamming threat. The path of the jammer is approximately from the top to the bottom of the figures. Frame (a) shows the primary network in the absence of a jammer. Frame (b) illustrates how the network re-structures as the jammer prevents

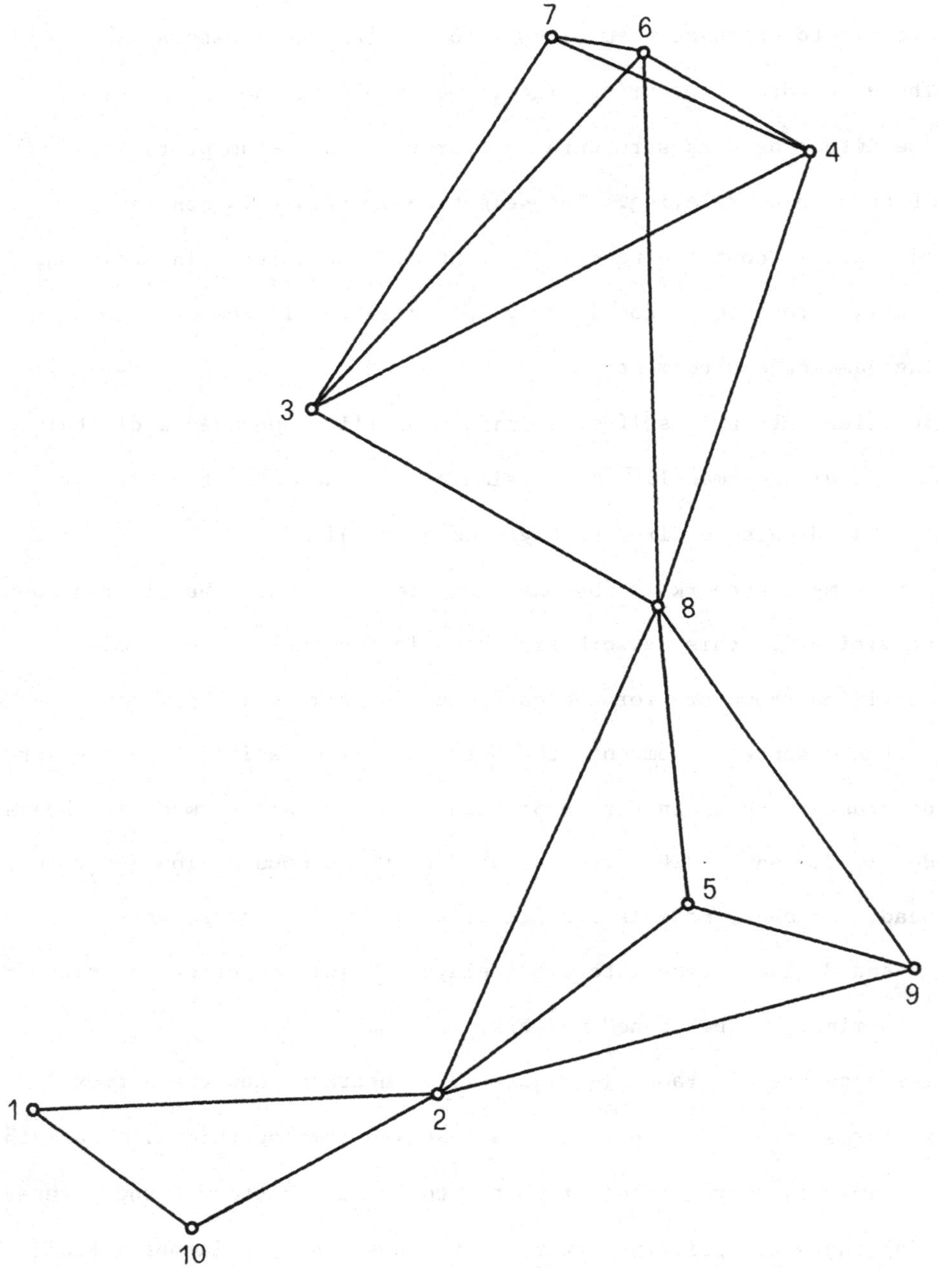

Fig. 12

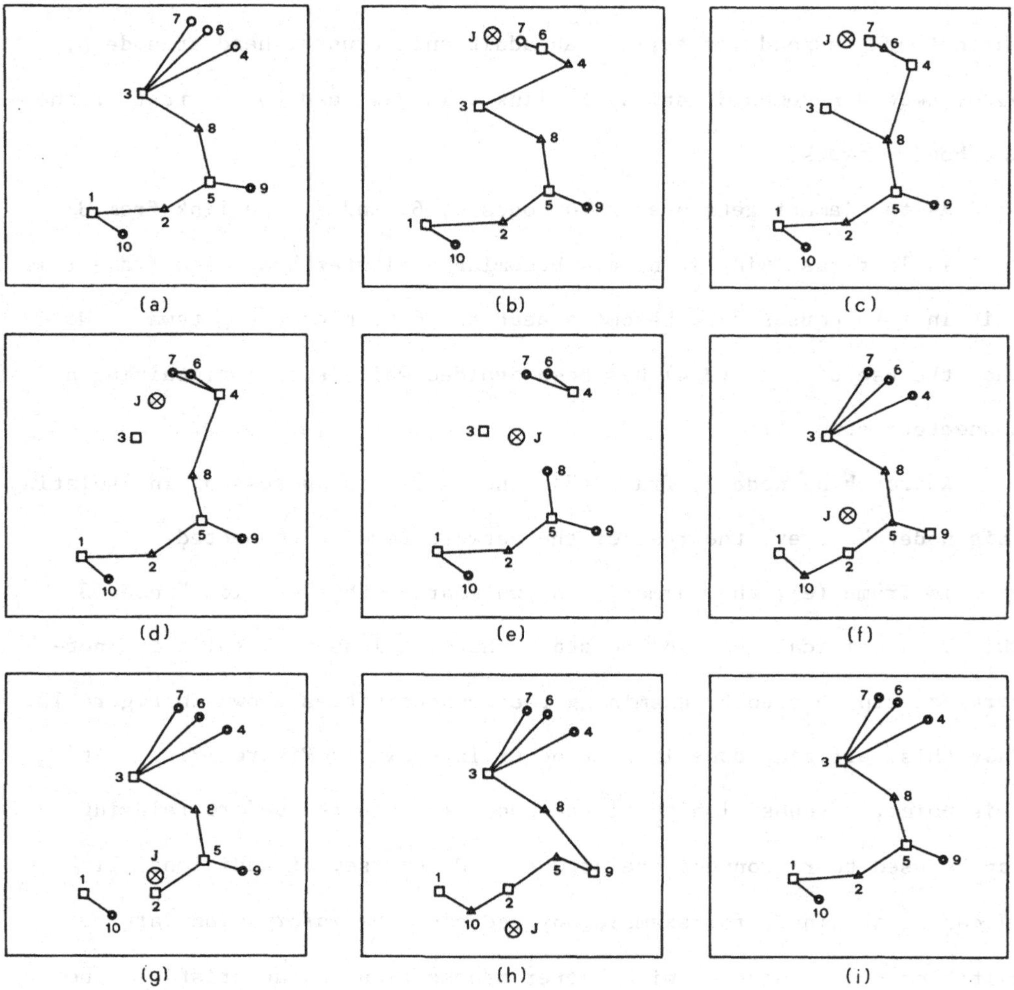

Fig. 13

node 7 from hearing node 3 and prevents node 6 from hearing either 3 or 8. Consequently, nodes 6 and 7 are no longer bidirectionally connected to cluster head 3 as they were in the case of Frame (a). The network responds by forming an additional cluster head at node 6; also, node 4 becomes a gateway to link this new head to the rest or the backbone network.

As the jammer gets nearer to nodes 4, 6, and 7, the link from 3 to 4 is lost resulting in node 4 becoming a cluster head, see frame (c). This in turn causes 7 to become a head and 6 to become a gateway. Note that the use of link (3,4) has been avoided while still maintaining a connected network.

Approaching node 3, frame (d), the jammer is successful in isolating this node, however, the rest of the network remains connected.

In frame (e), the jammer is shown near to the "critical" node 8. This is a critical node in the sense that its loss will split the network, as can be seen by examining the connectivities shown in Figure 12. That this splitting does in fact occur is shown in Figure 13.e. At this point, a repositioning of the nodes is required before relaying can be used to re-connect the network. Of course, if additional link AJ can be obtained, for example, by reducing the information rate or switching to a frequency with better propagation characteristics, then our network structuring algorithm will make use of the new links to form a connected network.

When the jammer moves away from nodes 3 and 8 to the position shown in (f), the network is able to return to the connected state. However, as the jammer approaches 2, the ability to use this node for relaying is lost and the network becomes disconnected again. As illustrated in frame (h), the network is able to recover to the connected state after a moderate separation between the jammer and node 2 occurs. Finally, with no jammer present, the network returns to its original configuration.

Thus we see that the basic algorithm described earlier allows the set of nodes to form an organized, connected network, without the use of any central controller or coordinator. Furthermore they allow the nodes to activate their "discovered" links in a distributed manner and maintain consistency in this activation. Finally we saw how by using the principle of relaying these same algorithms offer increased protection to interference by allowing the network to reconfigure itself in an adaptive, automatic fashion.

Conclusion

In these lectures we have examined certain special issues concerning security of digital radio communications amongst mobile users. After reviewing the fundamental aspects of multiple access we considered the often ignored aspects of capture and feedback and their potential impact on security and presented some new results. Following that we considered the problem of automatic, secure self-organization of the users into a network and displayed algorithms that achieve this goal while at the same time they provide inherent protection against interference.

Again I wish to express my thanks to CISM and to Professor G. Longo for their invitation that afforded me the opportunity to communicate these ideas to a distinguished audience.

References

1. Skwirzynski, J. (ed.), *Proceedings of NATO Advanced Study Institute on "New Concepts in Multi-user Communication"*, (Panel Discussion on Spread Spectrum for Mobile Communications), Sijthoff & Noordhoff, Alphen aan den Rijn, 1981.

2. Abramson, N., The aloha system - another alternative for computer communications , *AFIPS Conference Proceedings*, Fall Joint Computer Conference, Vol. 37, pp. 281-285, 1970.

3. Wieselthier, J. Ephremides, A., A new class of protocols for multiple access in satellite networks , *IEEE Trans. on AC*, p. 865, 1980.

4. Tobagi, F., Multiaccess protocols in packet communication systems , *IEEE Trans. Comm.*, Vol. 28, p. 468, 1980.

5. Fayolle, G., Gelenbe, E., Labetoulle, T., Stability and optimal control of the packet switching broadcast channel , *JACM*, Vol. 24, p. 375, 1977.

6. Lam, S.S., Kleinrock, L., Packet switching in a multiaccess broadcast channel: dynamic control procedures , *IEEE Trans. Comm.*, Vol. 23, p. 891, 1975.

7. Capetanakis, J., Tree algorithms for packet broadcast channels , *IEEE Trans. Info. Theory*, Vol. 25, p. 515, 1979.

8. Ephremides, A., On the capture effect in random access frequency hopped channels, *Proc. Princeton Conference on Information Sciences and Systems*, 1982.

9. Hajek, B., Recursive retransmission control-application to a frequency-hopped spread-spectrum system, *Proceedings Princeton Conference on Information Sciences and Systems*, 1982.

10. Abramson, N., Throughput of packet broadcasting channels, *IEEE Trans. on Communications*, 1977.

11. Metzner, On improving utilization in ALOHA networks, *IEEE Trans. on Communications*, 1976.

12. Rosner, R., *Packet Switching*, Lifetime Learning Publications, 1982.

13a. Cohn, D., Redundant packets in multiple access systems, *International Symposium on Information Theory*, Les Arcs, France, 1982.

13b. Snyder, D., *Random Point Processes*, Wiley, 1975.

14. Baker, D.J., Ephremides, A., The architectural organization of a mobile radio network via a distributed algorithm, *IEEE Trans. Comm.*, Vol. 29, p. 1694, 1981.

15. Baker, D.J., Wieselthier, J., Ephremides, A., A distributed algorithm for scheduling the activation of links in a self-organizing, mobil, radio network, *Proceedings ICC*, Philadelphia, 1982.

16. Cook, C.E., Optimum deployment of communications relays in an interference environment, *IEEE Trans. Comm.*, Vol. 28, p. 1608, 1980.

17. Cook, C.E., Anti-intercept margins of relay-augmented data links, *IEEE Trans. Comm.*, Vol. 29, p. 936, 1981.

18. Baker, D.J. Wieselthier, J.E., Ephremides, A., McGregor, D.N., Distributed network reconfiguration in response to jamming at HF, *MILCOM Proceedings*, Boston, 1982.

FAST DECODING ALGORITHMS FOR REED-SOLOMON CODES

R. E. Blahut

I. INTRODUCTION

Reed-Solomon codes and BCH codes of large blocklength and large alphabet size are coming into widespread use in communication and storage systems. Secure communication systems commonly use a Reed-Solomon code as one method of jam protection. The importance of these codes is partly due to their superior error-correcting performance, but it is equally due to the availability of efficient decoding algorithms for them. Future requirements will exist for ever-larger error-control codes, but only if the decoding cost remains affordable. It is important to reduce the complexity of the decoding algorithms. These lectures explore the complexity of such algorithms.

A number of efficient decoding algorithms for Reed-Solomon codes of blocklength n have been developed; decoders with on the order of n^2 computational steps. The purpose of these lectures is to sharpen the attack on the complexity of the decoders. One measure of computational complexity that we use is the number of multiplications in the computation. A secondary measure is the number of additions. In this sense, we will find techniques to reduce the computational complexity, but one pays a price for the improvement. To reduce the number of multiplications and additions, one uses a more intricate structure, and algorithms that contain more complicated indexing and branching. Of course, in many applications, the regularity of an algorithm is more important than the number of multiplications and additions. This aspect of complexity is difficult to quantify and is not treated.

II. DECODING OF REED-SOLOMON CODES

A Reed-Solomon code can be described using a discrete Fourier transform in the Galois field GF(q) and the language of digital signal processing. A Reed-Solomon code of blocklength n exists over GF(q) whenever GF(q) contains an nth root of unity. Similarly, a Fourier transform of blocklength n exists over GF(q) whenever GF(q) contains an nth root of unity.

Sometimes, to obtain a Fourier transform of blocklength n over GF(q), one must go into the extension field $GF(q^m)$ to find an nth root of unity. Then a Reed-Solomon code of blocklength n does not exist. It must be replaced by the more general code known as a BCH code.

The Fourier transform over any field is given by

$$V_j = \sum_{i=0}^{n-1} \omega^{ij} v_i \qquad j = 0, \ldots, n-1$$

where is an nth root of unity in the field. The Fourier transform over the complex field is familiar to many people. Over any other field it has pretty much the same properties, including the convolution theorem and the inverse Fourier transform. However, there need not be an nth root of unity for every n, so the Fourier transform will not exist for every n.

A Reed-Solomon code over GF(q) is a collection of vectors over GF(q) that are as mutually distinct as is possible. Specifically, let $\underline{c}$ be a vector over GF(q) with spectrum $\underline{C}$. A t-error-correcting Reed-Solomon code over GF(q) is the set of all n- vectors $\underline{c}$ whose spectrum satisfies $C_j = 0$ for $j = 1, \ldots, 2t$. One way to find these codewords is encode in the frequency domain. This means to insert q-ary symbols into the n-2t components of C_j that are not constrained to zero and then take the inverse Fourier transform. There are q^{n-2t} ways to do this and so there are q^{n-2t} codewords in the code.

The frequency-domain encoder is one way to map q-ary information vectors of length n-2t into codewords. Many other maps are popular. It does not matter which is used as long as it is a fixed one-to-one map. Then the information symbols can be recovered from the codewords.

The t-error-correcting Reed-Solomon code as we have defined it does indeed give a code that corrects t errors. The proof consists of describing a decoder that will correct t errors.

The codeward $\underline{c}$ is transmitted and the channel makes errors described by the vector $\underline{e}$ which is nonzero in not more than t places. The received word $\underline{v}$ is

$$v_i = c_i + e_i \qquad i = 0, \ldots, n-1$$

The decoder must process the received word so as to remove the error word $\underline{e}$; the information is then recovered from $\underline{c}$.

The syndromes of the noisy codeword $\underline{v}$ are defined by the following set of equations.

$$S_j = \sum_{i=0}^{n-1} \omega^{ij} v_i \qquad j = 1, \ldots, 2t$$

The syndromes obviously are 2t components of a Fourier transform. The received noisy codeword has Fourier transform components given by $V_j = C_j + E_j$ for $j = 0, \ldots, n-1$, and the syndromes are the 2t components of this spectrum from 1 to 2t. By construction of a Reed-Solomon code,

$$C_j = 0 \qquad j = 1, \ldots, 2t$$

Hence, the block of syndromes

$$S_j = V_j = E_j \qquad j = 1, \ldots, 2t$$

gives us a window through which we can look at 2t of the n components of the spectrum of the error pattern. The decoder must estimate the entire spectrum of the error pattern using the priori knowledge that at most t components of the error pattern are nonzero. The spectral estimator will be an autoregressive filter with taps described by a polynomial $\Lambda(x)$.

Suppose there are ν errors ($\nu \leq t$) at locations with index i_k for $k = 1, \ldots \nu$. Define the polynomial

$$\Lambda(x) = \prod_{k=1}^{\nu} (1 - x\omega^{i_k})$$

which is known as the error-locator polynomial. The vector Λ whose components Λ_j are coefficients of the polynomial $\Lambda(x)$ has inverse transform

$$\lambda_i = \frac{1}{n} \sum_{j=0}^{n-1} \Lambda_j \, \omega^{-ij}$$

which is just $\Lambda(x)$ evaluated at $x = \omega^{-i}$. But then

$$\lambda_i = \frac{1}{n} \prod_{k=1}^{\nu} (1 - \omega^{-i} \omega^{i_k})$$

Hence $\lambda_i = 0$ if and only if $e_i \neq 0$. This is why $\Lambda(x)$ is called the error-locator polynomial. Then $\lambda_i e_i = 0$ for all i, so the convolution in the frequency domain is zero

$$\Lambda * E = 0$$

but $\Lambda_j = 0$ for $j > t$ and $\Lambda_0 = 1$, so this can be written

$$\sum_{j=1}^{t} \Lambda_j E_{k-j} = -E_k \qquad k = 0, \ldots, n-1$$

This convolution is a set of n equations in n-t unknowns; the t unknown components of Λ and the n-2t unknown components of $\underline{E}$, and in the 2t known values of $\underline{E}$ given by the syndromes. Of the n equations, there are t equations that involve only components of Λ and the known components of $\underline{E}$ given by the syndromes. That is, the t equations

$$\sum_{j=1}^{t} \Lambda_j S_{k-j} = -S_k \qquad k = 1 + t, \ldots, 2t$$

involve only the known syndromes and the t unknown components of Λ. These t equations are always solvable for the t unknown components of Λ. The remaining components of $\underline{E}$ can be obtained by recursive extension. That is, sequentially compute from Λ using the above convolution equation written in the form

$$E_k = - \sum_{j=1}^{t} \Lambda_j E_{k-j} \qquad k = 0, \ldots, n-1$$

This equation describes the operation of an autoregressive filter with tap weights given by the coefficients of Λ. We can think of the decoding as the task of designing the autoregressive filter described by the polynomial $\Lambda(x)$; the filter then generates the missing components of $\underline{E}$. In this way all components of the vector $\underline{E}$ are computed. Finally,

$$C_j = V_j - E_j$$

and an inverse Fourier transform recovers the initial codeword with all errors corrected. The information symbols are read out in accord with the method of encoding.

III. THE BERLEKAMP-MASSEY ALGORITHM

The Berlekamp-Massey algorithm is a fast method for solving the Toeplitz system of linear equations

$$\sum_{j=1}^{t} S_{k-j}\Lambda_j = -S_k \qquad k = 1+t, \ldots, 2t$$

for the vector Λ. The complexity is proportional to t^2; it is an improvement over direct matrix inversion which has complexity proportional to t^3.

Solving this system of equations can be viewed as the task of designing an autoregressive filter to have a given output sequence; a task well-known in digital signal processing and arising in problems such as maximum-entropy spectral estimation or LPC speech compression.

The most popular algorithm for solving these equations for error-control decoders is the Berlekamp-Massey algorithm stated as follows:

Let $S_1, \ldots, S_{2t}$ be given. Let the following set of recursive equations be used to compute $\Lambda^{(2t)}(x)$:

$$\Delta_r = \sum_{j=0}^{n-1} \Lambda_j^{(r-1)} S_{r-j}$$

$$L_r = \delta_r (r-L_{r-1}) + (1-\delta_r)\, L_{r-1}$$

$$\begin{bmatrix} \Lambda^{(r)}(x) \\ B^{(r)}(x) \end{bmatrix} = \begin{bmatrix} 1 & -\Delta_r x \\ \Delta_r^{-1}\delta_r & (1-\delta_r)x \end{bmatrix} \begin{bmatrix} \Lambda^{(r-1)}(x) \\ B^{(r-1)}(x) \end{bmatrix}$$

$r = 1, \ldots, 2t$; the initial conditions are $\Lambda^{(0)}(x) = 1$, $B^{(0)}(x) = 1$, and $\delta_r = 1$ if both $\Delta_r \neq 0$ and $2L_{r-1} \leq r-1$ and otherwise $\delta_r = 0$. Then $\Lambda^{(2t)}(x)$ is the smallest-degree polynomial with the properties that $\Lambda_0^{(2t)} = 1$ and

$$S_k + \sum_{j=1}^{n-1} \Lambda_j^{(2t)} S_{k-j} = 0 \qquad k = L_{2t} + 1, \ldots, 2t$$

The Berlekamp-Massey algorithm can be used in any field but it seems not to have attracted much attention for problems in the real or complex field - perhaps because of the division by Δ which can be a small number.

IV. FAST CONVOLUTION ALGORITHMS

The best-known algorithms for cyclic convolutions uses the convolution theorem in conjunction with a fast Fourier transform. Many other algorithms for convolution, both linear and cyclic, are now known and often these are more efficient than using a fast Fourier transform and the convolution theorem. We will develop the Winogard algorithm for fast convolution in this section. The technique can be used in any field, but we are interested only in a finite field GF(q). The technique breaks a convolution into a number of short convolutions that are easy to compute. The short convolutions are recombined using the Chinese Remainder Theorem for polynomials.

The Winograd algorithm computes the remainder

$$c(x) = d(x)g(x) \qquad (\text{mod } b(x))$$

where b(x) is any fixed polynomial. To obtain an algorithm for the linear convolution

$$c(x) = d(x)g(x),$$

choose any integer N larger than deg c(x), and b(x) any polynomial of degree N. Then the remainder when c(x) is divided by b(x) equals c(x) and so the theory can also be used to compute linear convolutions. To break a convolution modulo b(x) into pieces, factor b(x) into relatively prime polynomials over some subfield of GF(q)

$$b(x) = b_1(x)b_2(x) \ldots b_s(x)$$

One would normally choose the prime subfield GF(p) for the factorization, so we study this case, but the theory can admit other subfields as well. The procedure will minimize the number of multiplications in GF(q), but will not attempt to minimize the number of multiplications in GF(p). In most applications to error control codes the prime field is GF(2) and those multiplications are trivial.

We divide the cyclic convolution into two steps. First compute the residues

$$c^{(k)}(x) = d(x)g(x) \qquad (\text{mod } b_k(x))$$

$$= d^{(k)}(x)g^{(k)}(x) \qquad (\text{mod } b_k(x))$$

for $k = 0, \ldots, s$ where the residues $c^{(k)}(x)$, $d^{(k)}(x)$ and $g^{(k)}(x)$ are the remainders when $c(x)$, $d(x)$ and $g(x)$ are divided by $b_k(x)$. Computation of the residues $d^{(k)}(x)$, $g^{(k)}(x)$ requires no multiplications.

By the Chinese Remainder Theorem for polynomials, $c(x)$ can be computed from this set of residues by

$$c(x) = a^{(0)}(x)c^{(0)}(x) + \ldots + a^{(s)}(x)c^{(s)}(x) \qquad (\text{mod } b(x))$$

for appropriate polynomials $a^{(0)}(x), \ldots, a^{(s)}(x)$, all with coefficients in GF(p). This last step requires no multiplications, only additions, because the polynomials $a^{(k)}(x)$ only have coefficients in the prime field.

The only multiplications in the big field GF(q) are in the short convolutions represented by the polynomial products $d^{(k)}(x)g^{(k)}(x)$; a total of $\sum_{k=1}^{s} [\deg b_k(x)]^2$ multiplications are required for the obvious implementation of these polynomial products because $d^{(k)}(x)$ and $g^{(k)}(x)$ each have $\deg b_k(x)$ coefficients. This can be a considerable savings in multiplications. Even further savings can be obtained in general by breaking one or more of the short convolutions down into even smaller pieces by repeating the same procedure.

The procedure can be made a little more efficient by choosing a $b(x)$ with a somewhat smaller degree. This will produce an incorrect convolution, but a few extra computations can correct it as follows. By the division algorithm, write

$$c(x) = Q(x)b(x) + R_{b(x)}[c(x)]$$

If deg $b(x) \leq$ deg $c(x)$, the Winograd algorithm will produce only the remainder. The term $Q(x)b(x)$ can be determined by a side computation and added in. The simplest case is the case where deg $b(x)$ = deg $c(x)$. Then $Q(x)$ must have degree zero. If $b(x)$ is a monic polynomial of degree N, then clearly $Q(x) = c_N$, the coefficient of x^N in the polynomial $c(x)$. Consequently

$$c(x) = c_N b(x) + R_{b(x)}\left[c(x)\right]$$

and c_N can be easily computed with one multiplication as the product of the leading coefficients of $d(x)$ and $g(x)$.

A cyclic convolution can be viewed as a polynomial product modulo x^n-1. The Winograd convolution algorithm can also be used here by choosing x^n-1 for $b(x)$. Now one must accept whatever prime factors x^n-1 happens to have; they cannot be chosen freely.

We will discuss a simple example, an encoder for the (7,3) Reed-Solomon code. This is only a small problem, so the power of the techniques will not be illustrated fully, but the development of the algorithm will be easy to follow.

A (7,3) Reed-Solomon code with generator polynomial $g(x) = x^4 + \alpha^3 x^3 + x^2 + \alpha x + \alpha^3$ has deg $d(x) \leq 2$. Direct computation of $g(x)d(x)$ requires fifteen multiplications. However for this particular $g(x)$, only six multiplications are needed, because $g(x)$ has only two distinct nonunity coefficients. The fast algorithm we derive next should be compared to both 15 and to 6 multiplications.

We can choose any $b(x)$ of degree not less than 7. We will choose

$$b(x) = x(x+1)(x^2+x+1)(x^3+x+1)$$
$$= b_0(x)b_1(x)b_2(x)b_3(x)$$

First we will find all of the fixed polynomials needed in the algorithm.

$$g^{(0)}(x) = R_{b_0(x)}[g(x)] = \alpha^3$$

$$g^{(1)}(x) = R_{b_1(x)}[g(x)] = \alpha$$

$$g^{(2)}(x) = R_{b_2(x)}[g(x)] = \alpha x+1$$

$$g^{(3)}(x) = R_{b_3(x)}[g(x)] = 1$$

From the data polynomial d(x), we compute the residues

$$d^{(0)}(x) = d_0$$

$$d^{(1)}(x) = d_2 + d_1 + d_0$$

$$d^{(2)}(x) = (d_1+d_2)x + (d_1+d_2)$$

$$d^{(3)}(x) = d_2x^2 + d_1x + d_0$$

These residues are computed without multiplications.

We now have four short convolutions to compute.

$$c^{(0)}(x) = g^{(0)}(x)d^{(0)}(x) = \alpha^3 d^{(0)}(x)$$

$$c^{(1)}(x) = g^{(1)}(x)d^{(1)}(x) = \alpha d^{(1)}(x)$$

$$c^{(2)}(x) = g^{(2)}(x)d^{(2)}(x) = (x+1)d^{(2)}(x) \pmod{x^2+x+1}$$

$$c^{(3)}(x) = g^{(3)}(x)d^{(3)}(x) = d^{(3)}(x)$$

To compute $c^{(0)}(x)$ requires one multiplication, to compute $c^{(1)}(x)$ requires one multiplication, to compute $c^{(2)}(x)$ requires two multiplications and to compute $c^{(3)}(x)$ requires none. By combining all of these pieces, the convolution can be written

$$c(x) = a^{(0)}(x)c^{(0)}(x) + \dots + a^{(3)}(x)c^{(3)}(x) \pmod{x^7+1}$$

where $a^{(i)}(x)$ for i=0, ..., 3 are polynomials, given by the Chinese Remainder Theorem, that have only 0 and 1 as coefficients. This step uses no multiplications. When computed in this way, only four multiplications in GF(8) are needed to compute the codeword c(x).

The Winograd convolution algorithm is convenient to use only when the blocklength is small. When the blocklength is large, an algorithm known as the Agarwal-Cooley algorithm should be used to break up the large convolution. For this to work, the convolutions must be viewed as cyclic convolutions. Good convolution algorithms will use the Agarwal-Cooley algorithm to break a long convolution into short convolutions combined with the Winograd algorithm to do the short convolutions efficiently.

Given the vectors d_i and g_i for i = 0, ..., n-1, we wish to compute the cyclic convolution

$$c_i = \sum_{k=0}^{n-1} g_{((i-k))} d_k \qquad i = 0, \dots, n-1$$

where the double parentheses on the indices designates modulo n.

We will turn this one-dimensional convolution into a two-dimensional convolution, by using the Chinese Remainder Theorm. Replace the indices i and k by double indices (i', i") and (k', k") given by

$$i' = i \quad (\text{mod } n')$$

$$i'' = i \quad (\text{mod } n'')$$

and

$$k' = k \quad (\text{mod } n')$$

$$k'' = k \quad (\text{mod } n'')$$

The Chinese Remainder Theorem tells how the original indices can be recovered from the new indices:

$$i = N''n''i' + N'n'i'' \quad (\text{mod } n)$$

$$k = N''n''k' + N'n'k'' \quad (\text{mod } n)$$

where N' and N" are those integers that satisfy

$$N'n' + N''n'' = 1 \quad (\text{mod } n)$$

The convolution

$$c_i = \sum_{k=0}^{n-1} g_{((i-k))} d_k$$

can be written

$$c_{N''n''i''N'n'i''} = \sum_{k'=0}^{n'-1} \sum_{k''=0}^{n''-1} g_{N''n''(i'-k') + N'n'(i''-k'')} \, d_{N''n''k'+N'n'k''}$$

The double summation on k',k" is equivalent to the single summation on k because it picks up the same terms. Now define the two-dimensional variables also called d, g, and c so that the convolution now becomes

$$c_{i',i''} = \sum_{k'=0}^{n'-1} \sum_{k''=0}^{n''-1} g_{i'-k',\, i''-k''} \, d_{k',k''}$$

where the first and second indices are interpreted modulo n' and modulo n'' respectively. This is now a two-dimensional cyclic convolution. However, there is not yet any improvement in computational complexity.

For the algorithm to be useful, we must give an efficient way to do the two-dimensional cyclic convolution. We use the Winograd short convolution algorithm along each axes. We will bring in the Winograd algorithm on only one axis to start; later it will be used on the other axis as well. This will be easiest to see if the two-dimensional arrays are now throught of as one-dimensional arrays of polynomials. Define the following vector of length n'' of polynomials of n'-1.

$$d_{k''}(x) = \sum_{k'=0}^{n'-1} d_{k',k''} x^{k'} \qquad k'' = 0, \ldots, n''-1$$

Similarly define the vectors of polynomials $g_{k''}(x)$, $c_{k''}(x)$. The two-dimensional convolution now becomes a one-dimensional convolution of polynomials.

$$c_{i''}(x) = \sum_{k''=0}^{n''-1} g_{((i''-k''))}(x) \, d_{k''}(x) \qquad (\text{mod } x^{n'}-1)$$

for i'' = 0, ..., n''-1. Of course, this is really the same calculation as before, but we are better able to think of the two dimensions in different ways. A Winograd cyclic convolution algorithm is an identity that works as well for a convolution of polynomials as for a convolution of numbers. Hence it can be used here. Each multiplication required of the Winograd algorithm is a multiplication of polynomials; that is, a convolution. It, in turn, is computed by another Winograd algorithm. In this way the two dimensional convolution is computed by nesting two Winograd convolution algorithms.

V. THE WINOGRAD FAST FOURIER TRANSFORM

We have seen that a Reed-Solomon code can be decoded by a Fourier transform, followed by inverting a Toeplitz system, followed by an inverse Fourier transform. Thus algorithms for the fast Fourier transform can be applied. The best-known FFT is the radix-two Cooley-Tukey FFT. Because the blocklength of Reed-Solomon codes is usually not a power of two, the radix-two FFT algorithms cannot be used. However mixed-radix Cooley-Tukey FFT algorithms can be used to advantage.

There are other FFT algorithms that will do better than the Cooley-Tukey FFT. The Winograd FFT is intrinsically a mixed-radix FFT. It is better than a mixed-radix Cooley-Tukey FFT.

The Winograd FFT can be best described as a small Winograd FFT and a large Winograd FFT. The first is a way of constructing highly optimized routines for small blocklengths - usually for blocklengths 2, 3, 4, 5, 7, 9, and 16. The large Winograd FFT is a method for binding these small pieces together to get a large FFT.

The small Winograd FFT uses the Rader prime algorithm (or a generalization) to change a Fourier transform into a convolution; then an efficient Winograd convolution algorithm is used for the convolution.

Let us construct a binary five-point Winograd FFT that will compute

$$V_j = \sum_{i=0}^{4} \alpha^{ij} v_i \qquad j = 0, \ldots, 4$$

This transform is in GF(16) since the smallest m for which 5 divides 2^m-1 is 4.

First use the Rader prime algorithm. The Rader prime algorithm can be used to compute the Fourier transform in any field GF(q) whenever the blocklength n is a prime. Because n is a prime, we can make use of the structure of GF(n). This prime field is not to be confused with GF(q) nor with any subfield of GF(q).

Choose a primitive element π in the field GF(p). Then each integer less than n can be expressed as a power of π. The Fourier transform in GF(q) can be written as follows with the zero frequency component and the zero time component treated specially.

$$V_0 = \sum_{i=0}^{n-1} v_i$$

$$V_j = v_0 + \sum_{i=1}^{n-1} \alpha^{ij} v_i$$

$$= V_0 + \sum_{i=1}^{4} (\alpha^{ij}-1) v_i \qquad j = 1, \ldots, 4$$

For each i from 1 to n-1, let r(i) be the unique integer from 1 to n-1 such that in GF(p), $\pi^{r(i)} = i$. The function r(i) is a permutation on $\{1,2,\ldots,n-1\}$. Then V_j can be written

$$V_{\pi^{r(j)}} = V_0 + \sum_{i=1}^{n-1} (\alpha^{\pi^{r(i)+r(j)}} - 1) v_{\pi^{r(i)}}$$

Since r(i) is a permutation we can set $\ell = r(j)$: set $k = n-1-r(i)$; and use k as the index of summation to get

$$V_{\pi^\ell} = V_0 + \sum_{k=1}^{n-1} (\alpha^{\pi^{\ell-k}} - 1) v_{\pi^{n-1-k}}$$

or

$$V'_\ell = V_0 + \sum_{k=1}^{n-1} (\alpha^{\pi^{\ell-k}} - 1) v'_k$$

where $V'_\ell = V_{\pi^\ell}$ and $V'_k = v_{\pi^{n-1-k}}$ are scrambled input and ouput data sequences. A more convenient form is obtained by substracting V_0 from both sides. Then

$$V'_\ell - V_0 = \sum_{k=0}^{n-1} (\alpha^{\pi^{\ell-k}} - 1) v'_k$$

This is now the equation of a cyclic convolution between $\{v'_k\}$ and $\{\alpha^{\pi^k} - 1\}$. By scrambling the input and output indices, we have turned the Fourier transform into a convolution. As it is written, the number of operations needed to implement the convolution is still on the order of n^2. However, the convolution may be computed by a fast convolution algorithm discussed previously.

First rewrite the five-point Fourier transform as

$$V_0 = \sum_{i=0}^{4} v_i$$

$$V_j = v_0 + \sum_{i=1}^{4} \alpha^{ij} v_i$$

$$= V_0 + \sum_{i=1}^{4} (\alpha^{ij}-1) v_i \qquad j = 1, \ldots, 4$$

and work only with the terms $\sum_{i=1}^{4} (\alpha^{ij}-1) v_i$. In GF(5), the element 2 is primitive, and so in GF(5) we have:

$$2^0 = 1 \qquad 2^0 = 1$$
$$2^1 = 2 \qquad 2^{-1} = 3$$
$$2^2 = 4 \qquad 2^{-2} = 4$$
$$2^3 = 3 \qquad 2^{-3} = 2$$

Hence:

$$V'_{\ell} - V_0 = \sum_{k=0}^{3} (\alpha^{2^{\ell-k}} - 1) v'_k$$

The summation is recognized as a four-point cyclic convolution. By identifying terms, we can write the Rader filter as a polynomial over GF(16).

$$g(x) = \alpha^8 x^3 + \alpha x^2 + \alpha^{14} x + \alpha^4$$

where $g_k = \alpha^{2^{-k}} - 1$. The input to the filter and the output from the filter can also be expressed as polynomials whose coefficients are scrambled coefficients of v and V.

$$a(x) = v_3x^3 + v_4x^2 + v_2x + v_1$$

$$b(x) = (V_2-V_0)x^3 + (V_4-V_0)x^2 + (V_3-V_0)x + V_1-V_0$$

and

$$b(x) = g(x)a(x) \qquad (\text{mod } x^4-1)$$

The polynomial g(x) is fixed. The polynomial a(x) is formed by scrambling the coefficients of v(x). The polynomial V(x) is obtained by unscrambling the coefficients of the polynomial b(x).

The five point small Winograd FFT is obtained if the product a(x)g(x) is computed by a small convolution algorithm. There is a four-point cyclic convolution algorithm with nine multiplications. We can rewrite this to do the Fourier transform. Incorporate the scrambling and unscrambling operations into the convolution by scrambling the appropriate rows and columns. Also the coefficients of g(x) are fixed constants in GF(16), so it is possible to precompute terms involving only g(x). When these changes are made to the four-point convolution algorithm, and the terms V_0 and v_0 are included, it becomes the five-point small Winograd FFT. It can be written in the form

$$V = BDAv$$

$$\begin{bmatrix} V_0 \\ V_1 \\ V_2 \\ V_3 \\ V_4 \end{bmatrix} = \begin{bmatrix} 1 & 0 & 0 & 0 & 0 & 0 & 0 & 0 & 0 & 0 \\ 0 & 1 & 0 & 1 & 1 & 0 & 0 & 0 & 0 & 1 \\ 0 & 1 & 1 & 0 & 0 & 1 & 1 & 0 & 0 & 0 \\ 0 & 1 & 0 & 1 & 0 & 1 & 0 & 0 & 1 & 0 \\ 0 & 1 & 1 & 0 & 1 & 0 & 0 & 1 & 0 & 0 \end{bmatrix} \begin{bmatrix} 1 & & & & & & & & & \\ & \alpha^{13} & & & & & & & & \\ & & \alpha^{9} & & & & & & & \\ & & & \alpha^{10} & & & & 0 & & \\ & & & & \alpha^{6} & & & & & \\ & & & & & 1 & & & & \\ & & & & & & \alpha^{4} & & & \\ & & 0 & & & & & \alpha^{14} & & \\ & & & & & & & & \alpha & \\ & & & & & & & & & \alpha^{8} \end{bmatrix} \begin{bmatrix} 1 & 1 & 1 & 1 & 1 \\ 0 & 1 & 0 & 0 & 0 \\ 0 & 1 & 0 & 0 & 1 \\ 0 & 1 & 0 & 0 & 1 \\ 0 & 1 & 0 & 1 & 0 \\ 0 & 1 & 1 & 0 & 0 \\ 0 & 1 & 1 & 1 & 1 \\ 0 & 1 & 1 & 1 & 1 \\ 0 & 1 & 1 & 1 & 1 \\ 0 & 1 & 1 & 1 & 1 \end{bmatrix} \begin{bmatrix} v_0 \\ v_1 \\ v_2 \\ v_3 \\ v_4 \end{bmatrix}$$

Notice that the matrix of preadditions A and the matrix of postadditions B are not square. The five-point input vector is expanded to a ten-point vector, and this is where the multiplications occur as represented by the diagonal matrix D. The top row inside the braces has to do with V_0; and has no multiplications. The other nine rows come from the four-point cyclic convolution algorithm. One of the multiplying constants turns out to be a one, so there are really only eight multiplications in the algorithm.

The small Winograd FFT of length n can be derived in this way whenever n is a prime. A small Winograd FFT also can be derived whenever n is a prime power, but we will not discuss the more general method. Generally, one constructs the small Winograd FFT only for fairly small blocklengths. For large blocklengths, one prefers to use something with a little more structure even if there is a slight penalty in the number of multiplications. The large Winograd FFT satisfies this need.

The large Winograd FFT has a blocklength n which is a product of small primes or small prime powers. We will discuss the case with two factors. Then $n = n'n''$. The first step is to change the one-dimensional Fourier transform into a two-dimensional Fourier transform using the Good-Thomas indexing scheme.

The Good-Thomas indexing scheme is based on the Chinese Remainder Theorem for integers. The input index is described by its residues as follows

$$i' = i \pmod{n'}$$

$$i'' = i \pmod{n''}$$

This is the map of the input index i down the extended diagonal of a two-dimensional array indexed by (i', i''). By the Chinese Remainder Theorem there exist integers N' and N'' such that

$$N'n' + N''n'' = 1 \pmod{n}$$

Then the input index can be recovered as follows:

$$i = i'N''n'' + i''N'n' \pmod{n}$$

The output index is described somewhat differently. Define

$$j' = N''j \pmod{n'}$$

$$j'' = N'j \pmod{n''}$$

These can be written in the equivalent form $j'' = ((N' \bmod n'')j \bmod n''))$ and $j' = ((N'' \bmod n')j \bmod n'))$. The output index j can be recovered as follows:

$$j = n''j' + n'j'' \pmod{n}$$

To verify this, write it out.

$$j = n''(N''j + Q_1 n') + n'(N'j + Q_2 n'') \pmod{n'n''}$$

$$= j(n''N'' + n'N') \pmod{n}$$

$$= j$$

Now, with these new indices, we convert the formula

$$V_j = \sum_{i=0}^{n-1} \alpha^{ij} v_i$$

into

$$V_{n''j' + n'j''} = \sum_{i''=0}^{n''-1} \sum_{i'=0}^{n'-1} \alpha^{(i'N''n''+i''N'n')(n''j'+n'j'')} v_{i'N''n''+i''N'n'}$$

Multiply out the exponent. Since α has order $n'n''$, terms in the exponent involving $n'n''$ can be dropped. Treat the input and output vectors as two-dimensional arrays using the index transformations given above. Then

$$V_{j',j''} = \sum_{i'=0}^{n'-1} \sum_{i''=0}^{n''-1} \alpha^{N''(n'')^2 i'j'} \alpha^{N'(n')^2 i''j''} v_{i',i''}$$

$$= \sum_{i'=0}^{n'-1} \sum_{i''=0}^{n''-1} \beta^{i'j'} \gamma^{i''j''} v_{i',i''}$$

where $\beta = \alpha^{N''(n'')^2}$ is an element of order n' and $\gamma = \alpha^{N'(n')^2}$ is an element of order n''. This is now in the form of a two-dimensional n' by n'' point Fourier transform.

The individual components of this two-dimensional Fourier transform can be computed by an n'-point Winograd FFT and an n"-point Winograd FFT respectively. This consists of taking an n'-point Fourier transform of each row followed by an n"-point Fourier transform of each column.

There is yet one more step before we have the large Winograd FFT. Since it does not matter whether the rows or the columns of the two-dimensional Fourier transform are transformed first, it seems that it may be possible somehow to do them together. This is what the Winograd FFT does. It binds together the row computations and the column computations in a way that reduces the total number of multiplications. The technique uses the notion of a Kronecker product of matrices.

Let $A = (a^{ik})$ be an I by K matrix and let $B = (b_{j\ell})$ be a J by L matrix. Then the Kronecker product of A and B, denoted A x B, is a matrix with IJ rows and KL columns whose entry in row (i-1)J+j and column $(k-1)L+\ell$ is given by

$$c_{ij,k\ell} = a_{ik} b_{j\ell}$$

The Kronecker product satisfies (A x B)(C x D) = (AC) x (BD) provided the matrix products all exist. To verify this, let the matrices A, B, C, and D have respective dimensions I x K, J x L, K x M and L x N. Since A x B has KL columns and C x D has KL rows, the matrix product (A x B)(C x D) is defined. It has IJ rows which we doubly index by (i,j) and MN columns which we doubly index by (m,n). The entry in row (i,j) and column (m,n) is $\Sigma_{k\ell} a_{ik} b_{j\ell} c_{km} d_{\ell n}$. Since AC has I rows and M columns, and BD has J rows and L columns, (AC)x(BD) also is an IJ by MN matrix. Its entry in row (i,j) and column (m,n) is

$$\Sigma_k a_{ik} c_{km} \Sigma_\ell b_{j\ell} d_{\ell n} = \Sigma_{k\ell} a_{ik} b_{j\ell} c_{km} d_{\ell n}$$

as we asserted.

The Kronecker product has application to the Fourier transform. Let W' and W" be matrix representations of Fourier transforms of size n' and n" respectively. This is:

$$V' = W'v'$$

$$V'' = W''v$$

are matrix representations of the Fourier transforms

$$V'_j = \sum_{i=0}^{n'-1} \beta^{ij} v'_i$$

$$V''_j = \sum_{i=0}^{n''-1} \gamma^{ij} v''_i$$

An n' by n" two-dimensional Fourier transform of the two-dimensional signal $v_{i'i''}$ is obtained by applying W' to each row, and then applying W" to each column. An n' by n" two-dimensional signal $v_{i'i''}$ can be turned into a one-dimensional signal by reading it by rows. (The two-dimensional signal was previously formed from a one-dimensional signal by the Chinese Remainder Theorem. Reading it by rows results in a new one-dimensional signal whose components are a permutation of the components of the original one-dimensional signal.)

If we think of $v_{i'i''}$ and $V_{j'j''}$ arranged as one-dimensional n'n" point vectors in this way, then the two-dimensional transform can be written using a Kronecker product

$$V = (W' \times W'')v$$

Whenever we have a Winograd transform of length n' and n", then we have the matrix factorizations

$$W' = B'D'A'$$

$$W'' = B''D''A''$$

where A', A", B' and B" are matrices of integers of the field; and D' and D" are diagonal matrices with elements from GF(q). The multiplication by matrix D' or D" is where the Winograd algorithm collects all of its multiplications. Let W = W' x W", and apply the theorem twice to get:

$$\begin{aligned} W &= (B'D'A') \times (B''D''A'') \\ &= (B' \times B'')(D' \times D'')(A' \times A'') \\ &= BDA \end{aligned}$$

where the Kronecker products B = B'xB" and A = A'xA" have only elements from GF(p), and the Kronecker product D = D'xD" is again a diagonal matrix with elements from GF(q). Hence, we have an n'n" Fourier transform algorithm, again in the form of the Winograd FFT. In this way large Winograd FFT algorithms can be built up from small ones. However, we have derived this algorithm with the presumption that v is in a scrambled order and the algorithm computes V in a scrambled order. However, once A and B are derived, it is trivial to rearrange the columns of A and the rows of B so that v and V are in their natural order.

VI. A FAST BERLEKAMP-MASSEY ALGORITHM

The Berlekamp-Massey algorithm requires a number of multiplications in the rth iteration which is approximately equal to twice the degree of $\Lambda^{(r)}(x)$. Since deg $\Lambda^{(r)}(x)$ is on the order of r, typically $\frac{r}{2}$, and there are 2t iterations, about $\sum_{r=0}^{2t} r$ or $2t^2$ multiplications are required and about the same number of additions. This is quadratic in t, so there are $O(t^2)$ multiplications in the Berlekamp-Massey algorithm. For very large codes and large t, the number of multiplications can be a burden. In this section, we will show a way to reduce the computational complexity for long codes.

The fast algorithm can be derived as a doubling algorithm. Divide the number of iterations in half. Modify the equations so that each half can be solved separately, and the two half solutions merged into the desired solution. If this works, and the form of the equations is unchanged, then the same idea can be applied to each of the two halves. Thus, a recursive algorithm is obtained.

The development will begin with a more compact organization of the Berlekamp-Massey algorithm. We replace the polynomials $\Lambda^{(r)}(x)$ and $B^{(r)}(x)$ by a 2 by 2 matrix of polynomials.

$$\underset{\sim}{\Lambda}^{(r)}(x) = \begin{bmatrix} \Lambda_{11}^{(r)}(x) & \Lambda_{12}^{(r)}(x) \\ & \\ \Lambda_{21}^{(r)}(x) & \Lambda_{22}^{(r)}(x) \end{bmatrix}$$

The element $\Lambda_{ab}^{(r)}(x)$ is a polynomial with coefficients denoted by $\Lambda_{ab,j}^{(r)}$. The matrix $\underset{\sim}{\Lambda}^{(r)}(x)$ will be defined in such a way that $\Lambda^{(r)}(x)$ and $B^{(r)}(x)$ can be computed by the equations

$$\Lambda^{(r)}(x) = \Lambda_{11}^{(r)}(x) + \Lambda_{12}^{(r)}(x)$$

$$B^{(r)}(x) = \Lambda_{21}^{(r)}(x) + \Lambda_{22}^{(r)}(x)$$

Recall that the computations of the Berlekamp-Massey algorithm reside primarily in the two equations

$$\Delta_r = \sum_{j=0}^{n-1} \Lambda_j^{(r-1)} S_{r-j}$$

$$\begin{bmatrix} \Lambda^{(r)}(x) \\ B^{(r)}(x) \end{bmatrix} = \begin{bmatrix} 1 & -\Delta_r x \\ \Delta_r^{-1}\delta_r & (1-\delta_r)x \end{bmatrix} \begin{bmatrix} \Lambda^{(r-1)}(x) \\ B^{(r-1)}(x) \end{bmatrix}$$

The second equation can be expanded.

$$\begin{bmatrix} \Lambda^{(r)}(x) \\ B^{(r)}(x) \end{bmatrix} = \begin{bmatrix} 1 & -\Delta_r x \\ \Delta_r^{-1}\delta_r & (1-\delta_r)x \end{bmatrix} \cdots \begin{bmatrix} 1 & -\Delta_1 x \\ \Delta_1^{-1}\delta_1 & (1-\delta_1)x \end{bmatrix} \begin{bmatrix} 1 \\ 1 \end{bmatrix}$$

Define the matrix $\underset{\sim}{\Lambda}^{(r)}(x)$ by

$$\underset{\sim}{\Lambda}^{(r)}(x) = \begin{bmatrix} 1 & -\Delta_r x \\ \Delta_r^{-1}\delta_r & (1-\delta_r)x \end{bmatrix} \cdots \begin{bmatrix} 1 & -\Delta_1 x \\ \Delta_1^{-1}\delta_1 & (1-\delta_1)x \end{bmatrix}$$

From this matrix $\Lambda^{(r)}(x)$ and $B^{(r)}(x)$ can be obtained by the expression

$$\begin{bmatrix} \Lambda^{(r)}(x) \\ B^{(r)}(x) \end{bmatrix} = \underset{\sim}{\Lambda}^{(r)}(x) \begin{bmatrix} 1 \\ 1 \end{bmatrix}$$

It serves just as well to update $\underset{\sim}{\Lambda}^{(r)}(x)$ as to update $\Lambda^{(r)}(x)$ and $B^{(r)}(x)$, although updating $\underset{\sim}{\Lambda}^{(r)}(x)$ directly can involve about twice as many multiplications because it has four elements rather than two. We will replace the iterates $\Lambda^{(r)}(x)$ and $B^{(r)}(x)$ by the iterate $\underset{\sim}{\Lambda}^{(r)}(x)$ and accept the doubling of the number of multiplications. This penalty will be overcome later by reorganizing the computations.

The recursive form of the Berlekamp-Massey algorithm is built around the two equations

$$\Delta_r = \sum_{j=0}^{n-1} \Lambda_{11,\,j}^{(r-1)} S_{r-j} + \sum_{j=0}^{n-1} \Lambda_{12,\,j}^{(r-1)} S_{r-j}$$

$$\underset{\sim}{\Lambda}^{(r)}(x) = \begin{bmatrix} 1 & -\Delta_r x \\ \Delta_r^{-1}\delta_r & (1-\delta_r)x \end{bmatrix} \underset{\sim}{\Lambda}^{(r-1)}(x)$$

To split the algorithm into halves, let

$$M^{(2t)}(x) = M'^{(t)}(x)\, M''^{(t)}(x)$$

where

$$M'^{(t)}(x) = \prod_{r=2t}^{t+1} \begin{bmatrix} 1 & -\Delta_r x \\ \Delta_r^{-1}\delta_r & (1-\delta_r)x \end{bmatrix}$$

$$M''^{(t)}(x) = \prod_{r=t}^{1} \begin{bmatrix} 1 & -\Delta_r x \\ \Delta_r^{-1}\delta_r & (1-\delta_r)x \end{bmatrix}$$

We will compute the two halves separately and then multiply them together. This will entail less work than the original organization.

We will also need to reorganize the equations for Δ_r. Think of Δ_r as the rth coefficient of the first component of the two-vector of polynomials.

$$\begin{bmatrix} \Delta(x) \\ \Delta'(x) \end{bmatrix} = \begin{bmatrix} \Lambda_{11}^{(r-1)}(x) & \Lambda_{12}^{(r-1)}(x) \\ \Lambda_{21}^{(r-1)}(x) & \Lambda_{22}^{(r-1)}(x) \end{bmatrix} \begin{bmatrix} S(x) \\ S(x) \end{bmatrix}$$

Hence for r larger than t

$$\begin{bmatrix} \Delta(x) \\ \Delta'(x) \end{bmatrix} = M'^{(r-1)}(x) M''^{(t)}(x) \begin{bmatrix} S(x) \\ S(x) \end{bmatrix}$$

$$= M'^{(r-1)}(x)\ S^{(t)}(s)$$

where

$$S^{(t)}(x) = M''^{(t)}(x) \begin{bmatrix} S(x) \\ S(x) \end{bmatrix}$$

This completes the splitting of the Berlekamp-Massey algorithm. The basic algorithm now is written

$$\Delta_r = \sum_{j=0}^{n-1} M_{11,j}^{(r-1)} S_{1,r-j} + \sum_{j=0}^{n-1} M_{12,j}^{(r-1)} S_{2,r-j}$$

$$M^{(r)}(x) = \begin{bmatrix} 1 & -\Delta_r x \\ \Delta_r^{-1}\delta_r & (1-\delta_r)x \end{bmatrix} M^{(r-1)}(x)$$

where $M^{(r)}(x)$ may represent either $M'^{(r)}(x)$ or $M''^{(r)}(x)$, and $(S_1(x), S_2(x))$ represent $(S(x), S(x))$ in the first half of the computation, and is updated to $S^{(t)}(x)$ in the second half. After both halves are complete, $M^{(2t)}(x)$ is obtained by multiplying its two halves.

The Berlekamp-Massey algorithm is now split into two halves. Notice that each of the halves itself is a Berlekamp-Massey algorithm. Hence if t is even the two halves can in turn be split: if t is a power of two, the splitting can continue until pieces that are only one iteration long are reached. These are executed, but are quite trivial.

The multiplication load of the recursive procedure is almost entirely in the convolutions. This is because when an iteration of the Berlekamp-Massey algorithm is finally executed, it occurs as a single iteration and has just one multiplication (by Δ_m^{-1}), which can be neglected when counting multiplications. The computational complexity of the recursive procedure depends directly on the computational complexity of the convolution algorithms. Its successful use presumes the availability of a good set of convolution algorithms.

We close the section with an informal discussion of the asymptotic computational complexity of the decoding of BCH codes. The general case of a decoder requiring 2t iterations has a computational complexity as measured by the number of multiplications that is on the order of $(\log_2 2t)C(t)$ where $C(t)$ is the number of multiplications required to convolve polynomials of degree t. This is because there are 2^ℓ visits to level ℓ and level ℓ requires convolutions of length $2^{-\ell}t$. The computational complexity of 2^{ℓ} convolutions each of length $2^{-\ell}t$ is greatest when ℓ equals zero, and ℓ takes on $\log_2 2t$ values, so the computational complexity is on the order of $(\log_2 2t)$.

A bound on the asymptotic computational complexity of the recursive decoder depends on a bound on the asymptotic complexity of convolution in a Gaolis field. Recall that the Winograd convolutions require one to choose a set of relatively prime factors over GF(p). To minimize the number of multiplications, these should be chosen to have small degree. However, if n is large there will not be enough polynomials of small degree. Some polynomials whose degree is on the order of log n will be needed. The convolution of size n will be replaced by a number of convolutions, the largest of which has size on the order of log n. The same process can be used in turn to replace convolutions of size log n by a number of smaller convolutions, the largest of which has size about log (log n). By formalizing this argument, one can say that the asymptotic complexity of convolution in a Galois field is $O(n2^{\log^* n})$ where $\log^* n$ is the number of times that a base two logarithm must be iterated starting with argument n to get a number less than or equal to one. That is

$$\log_2(\log_2(\ldots\log_2 n) \leq 1$$

and $\log^* n$ is the number of iterations. This says that C(n) is at most $O(n2^{\log^* n})$ so that the recursive Berlekamp-Massey algorithm has complexity $O(n2^{\log^* n}\log n)$. The term $2^{\log^* n}$ goes to infinity more slowly than log(log(... log n) for any finite number of iterated logs. Hence, the recursive Berlekamp-Massey algorithm has complexity greater than O(nlogn) by the thinnest of margins.

VII. ACCELERATED DECODING OF BCH CODES

The techniques of the previous sections are now applied to obtain an accelerated decoder. It suffices to discuss only primitive Reed-Solomon codes. BCH codes and nonprimitive Reed-Solomon codes are decoded in the same way. A primitive t-error-correcting Reed-Solomon code of blocklength $n = q-1$ is the set of all words of length n over GF(q) whose spectrum is zero in a specified block of 2t consecutive components.

The decoder works with the Fourier transform of v, the received word, and corrects all error patterns of weight at most t. Fast algorithms for the Fourier transform have been discussed.

The codeword spectral component C_j is zero on a block of 2t components so we have the 2t syndromes:

$$S_j = E_j = V_j \qquad j = 1, \ldots, 2t$$

The error-locator polynomial

$$\Lambda(x) = \prod_{k=1}^{\nu} (1-x\alpha^{i_k})$$

where $\nu \leq t$ is the number of errors, satisfies

$$\sum_{j=0}^{t} \Lambda_j E_{k-j} = 0$$

The convolution is a set of n equations in n-t unknowns. Of the n equations, there are t equations that involve only components of Λ and known components of E. The recursive Berlekamp-Massey algorithm of the previous section will solve these equations for the unknown components of Λ with complexity at most $0(n\log^2 n)$. The remaining components of E can be recursively computed from Λ and the known components of E by

$$E_j = -\sum_{k=1}^{t} \Lambda_k E_{j-k}$$

The computation of E_j for $j = 2t+1, \ldots, n$ by the above equation requires $t(n-2t)$ multiplications and therefore has complexity of order n^2. In the remainder of this section, we show how to reduce the computational complexity of this step to complexity at most $0(n\log n)$.

The form of the Berlekamp-Massey algorithm given by Massey treats the spectrum E_j on an interval of length n, and as periodically continued outside of this interval. The relationship between E(x) and $\Lambda(x)$ can be written

$$E(x)\,\Lambda(x) = 0 \qquad \text{mod } x^n-1$$

After a Fourier transform, this becomes

$$e_i\lambda_i = 0 \qquad i = 0, \ldots, n-1$$

These time-domain equations are indeterminate since the solution is $e_i = 0/\lambda_i$ and $\lambda_i = 0$ whenever $e_i \neq 0$.

Berlekamp takes a slightly different approach to the algorithm. With the Berlekamp point of view, the shift register output is periodic only after a start-up transient. The shift register is initially empty and the known polynomial Ω is shifted in. The equation becomes

$$E_k = -\sum_{j=1}^{t} \Lambda_j E_{k-j} + \Omega_k$$

and the excitation Ω_k, zero for $k > t$, brings the shift register from the initially quiescent state into the periodic state described previously .

The error spectrum polynomial E(x) satisfies the linear convolution

$$E(x)\ \Lambda\ (x) = \Omega(x)$$

The polynomial $\Omega(x)$ can be calclulated by expanding the recursive algorithm to include it. This requires no asymptotic increase in complexity, and E(x) can be computed by polynomial division. The polynomial division also has complexity that grows as n^2. Because the expression is neither periodic nor of finite duration, transform techniques are not immediately applicable. To solve this equation, however, the Forney algorithm is an indirect method that does the job. From $\Lambda(x)$ compute $\Omega(x)$ by

$$\Omega(x) = E(x)\,\Lambda\,(x) \qquad \bmod x^{2t}$$

and compute $\Lambda'(x)$ as the formal derivative of $\Lambda(x)$. Then take inverse Fourier transforms of Λ, Ω, and Λ' to get vectors λ_i, ω_i, and λ'_i. The error value is given by

$$e_i = -\frac{\omega_i}{\alpha^i \lambda'_i}$$

whenever λ_i is equal to zero. Since the Forney algorithm can be done with three Fourier transforms and a convolution, it has complexity O(nlogn).

References

1. Cooley, J.W. and J.W. Tukey, "An Algorithm for the Machine Computation of Complex Fourier Series," Math. Comp. Vol. 19, pp. 297-301, 1965.

2. Good, I.J., "The Interaction Algorithm and Practical Fourier Analysis," J. Royal Statist. Soc., Ser. B Vol. 20, pp. 361-375, 1958; addendum, Vol. 22, pp. 372-375, 1960.

3. Thomas, L.H., "Using a Computer to Solve Problems in Physics," in Applications of Digital Computers, Boston, Mass., Ginn and Co., 1963.

4. Justesen, J., "On the Complexity of Decoding Reed-Solomon Codes,: IEEE Trans. on Information Theory, Vol. IT-22, pp. 237-238, 1976.

5. Sarwate, D.V., "On the Compexity of Decoding Goppa Codes," IEEE Trans. on Information Theory Vol. IT-23, pp. 515-516, 1977.

6. Agarwal, R., and J.W. Cooley, "Algorithms for Digital Convolution," IEEE Trans. on Acoustics, Speech and Signal Processing, Vol. ASSP-25, pp. 392-410, 1977.

7. Winograd, S., "On Computing the Discrete Fourier Transform," Math. Comp. Vol. 32, pp. 175-199, 1978.

8. Nussbaumer, H.J. Fast Fourier Transform and Convolution Algorithms, Springer Verlag, Berlin, 1981.

9. Miller, R.L., T.K. Truong, and I.S. Reed, "Efficient Program for Decoding the (255, 223) Reed-Solomon Code Over GF(2^8) with Both Errors and Erasures, Using Transform Decoding," IEEE Proceedings, Volume 127, pp. 136-142, 1980.

10. Blahut, R.E., "Efficient Decoder Algorithms Based on Spectral Techniques," IEEE Abstracts of Papers - IEEE International Symposium on Information Theory, Santa Monica, California, 1981.

11. Preparata, F.P., and D.V. Sarwate, "Computational Complexity of Fourier Transforms over Finite Field," Mathematics of Computation, Vol. 31, pp. 740-751, 1977.

PSEUDO-RANDOM SEQUENCES WITH A PRIORI DISTRIBUTION*

Joshua H. Rabinowitz
The MITRE Corporation
Bedford, MA 01730
U.S.A.

This paper introduces a hypergraph-theoretic technique for the synthesis of sequences with prescribed distributions. A positive n-distribution is an assignment of a positive integer to each n-letter word with letters chosen from some fixed finite alphabet. A distribution is atomic if it always assigns the same integer to words that are cyclic shifts of each other. A periodic sequence of letters from the alphabet is said to realize a given distribution if the distribution assigns to each word its frequency of occurrence in a single period of the sequence. Given a positive atomic n-distribution, we construct a hypergraph, the subgraphs of which correspond to a family of modified nonlinear feedback shift registers. Each shift register in a family corresponding to a spanning hypertree of the hypergraph will generate a sequence realizing the given distribution.

*This research was accomplished as part of the MITRE Corporation's Independent Research and Development Program.

1. Introduction and Summary of Results

This paper introduces a hypergraph-theoretic technique for the synthesis of pseudo-random sequences with prescribed distributions. The technique is a generalization of a method devised by the author for the synthesis of k-ary de Bruijn sequences.

Two graphical techniques for the synthesis of binary de Bruijn sequences exist.[1,2] The first employs a rooted directed spanning tree in the binary de Bruijn graph of span n to generate a binary de Bruijn sequence of span n. This technique was introduced by de Bruijn for the enumeration of de Bruijn sequences, and it can be used to generate all such sequences. The second technique uses an undirected spanning tree in the binary necklace graph. This method generates only sequences with minimal-weight truth tables. Both techniques have been generalized to k-ary de Bruijn sequences, the first by von Aardenne-Ehrenfest and de Bruijn[3] and the second by the author.[4] This latter approach is generalized here to a synthesis technique for a large class of periodic sequences including the de Bruijn sequences.

Let L be the finite alphabet $L = \{0,...,k-1\}$ and let L^n be the ordered n-tuples of elements of L. An atom is an equivalence class of elements of L^n, where $v \sim w$ if v is a cyclic shift of w. A positive atomic n-distribution over L is a function d: $L^n \rightarrow N$, N the natural numbers, which is constant on atoms. A periodic sequence of elements of L has distribution d if each $v \in L^n$ appears precisely d(v) times in each period of the sequence. For each d, we construct a hypergraph P_d, each subgraph of which corresponds to a family of modified nonlinear feedback shift registers (NLFSRs) of span n over L. If T is a spanning hypertree of P_d, then each shift register in its corresponding family generates a periodic sequence with distribution d.

The paper is structured as follows: Section 2 recalls the relevant known results for the binary and k-ary de Bruijn sequence case. Section 3 is devoted to basic definitions and examples, and section 4 contains the principal results of the paper.

2. de Bruijn Sequences and Necklace Graphs

A **binary de Bruijn sequence** of span n is a periodic binary sequence (a_i) of period 2^n such that the n-tuples $S_i = (a_i,\ldots,a_{i+n-1})$ for $i=1,2,\ldots,2^n$ are distinct. For example, $\overline{00011101}$ is a de Bruijn sequence of span 3. (The bar indicates that the sequence is to be repeated indefinitely.) We wish to synthesize the sequences by finite-state machines with minimal memory by using an NLFSR of span n (figure 1). We

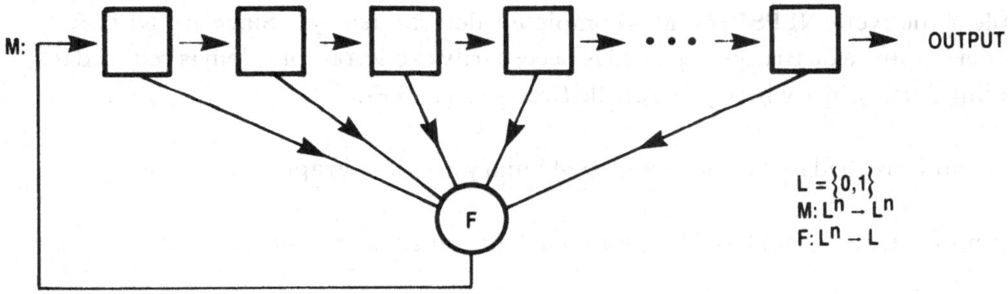

Figure 1. A Nonlinear Feedback Shift Register

restrict our attention to NLFSRs where next-state mappings are given by equations of the form

$$M(x_{n-1},\ldots,x_0) = (F(x_{n-1},\ldots,x_0), x_{n-1},\ldots,x_1)$$

and which output the rightmost bit x_0. Such a shift register is completely determined by the feedback function, F. An NLFSR is **cyclic** if the mapping $M:L^n \rightarrow L^n$ is one-to-one. M is cyclic if and only if there is a boolean function $f:L^n \rightarrow L$ such that

$$F(x_{n-1},\ldots,x_0) = f(x_{n-1},\ldots,x_1) + x_0$$

where + is addition mod 2 (see figure 2).[5] This function, f, is often called the **truth**

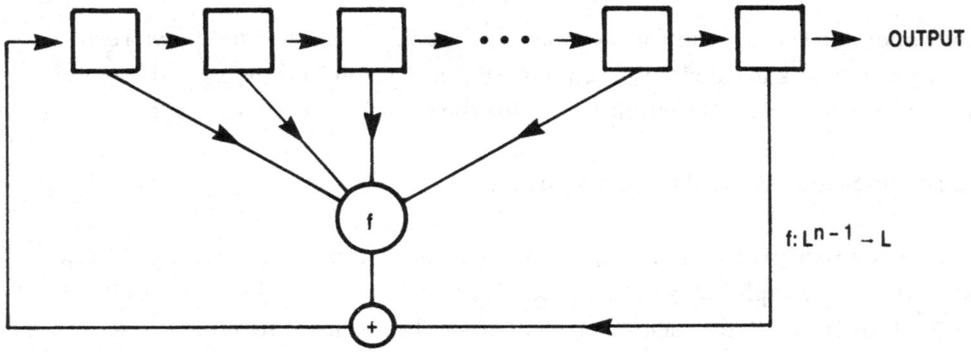

Figure 2. A Cyclic Nonlinear Feedback Shift Register

table of the cyclic NLFSR, M. M is completely determined by f. Since any NLFSR which outputs a de Bruijn sequence is necessarily cyclic, our problem is reduced to finding truth tables which generate de Bruijn sequences.

We now proceed to the construction of binary necklace graphs.

Definition: Let L = {0,1} and let n be a fixed integer greater than 1.

(a) A **state** is an ordered n-tuple of elements of L, i.e., an element of L^n.

(b) An **atom** or **necklace** is an equivalence class of states where v ~ w if v is a cyclic shift of w.

(c) A **link** is an element of L^{n-1}.

Roman letters denote states and Greek letters denote links. If v is a state then we denote by (v) the corresponding atom. If α is a link, then $\alpha \mid 1$ is a state and $(\alpha \mid 1)$ an atom, where $\mid$ denotes juxtaposition.

Definition: The **binary necklace graph** of span n, P_n, is given as follows:

(a) The **vertices** of P_n, $V(P_n)$, are the atoms.

(b) The **edges** of P_n, $E(P_n)$, are the links. The link α connects the vertices $(\alpha \mid 0)$ and $(\alpha \mid 1)$.

Proposition: There is a natural one-to-one correspondence between edge subgraphs of P_n and cyclic NLFSRs of span n.

Proof: Recall that an **edge subgraph** of a graph $G = (V,E)$ is a graph $G' = (V',E')$ with $V' = V$ and $E' \subset E$. Such a subgraph can be used to define a cyclic NLFSR. Let $G = (V(P_n),E')$ be a subgraph of P_n so that $E' \subset E\ (P_n)$. Since cyclic NLFSRs are completely determined by their truth tables, it suffices to define a truth table based on G. Now $E(P_n) \cong L^{n-1}$, and $E' \subset E(P_n)$, so E' may be viewed as a subset of L^{n-1}. Let $f: L^{n-1} \rightarrow L$ be given by $f(\alpha) = 0$ if $\alpha \notin E'$, and $f(\alpha) = 1$ if $\alpha \in E'$. f is a truth table. Hence, f determines a cyclic shift register as desired. Moreover, all cyclic NLFSRs can be obtained in this manner, and different subgraphs will define different shift registers, Q.E.D.

Definition: A subgraph of P_n is **distinguished** or a **d-subgraph** if the corresponding NLFSR outputs a de Bruijn sequence.

Definition: A **spanning tree** of a graph G is a subgraph of G that is a tree and contains all the vertices of G.

Theorem: Every spanning tree of P_n is distinguished and hence defines a de Bruijn sequence.[2,4]

As a simple example, consider the spanning tree T of P_4 in figure 3. This defines an NLFSR with the values of the truth table shown in table 1. The shift register, in turn, outputs the sequence with period 1001111010110000.

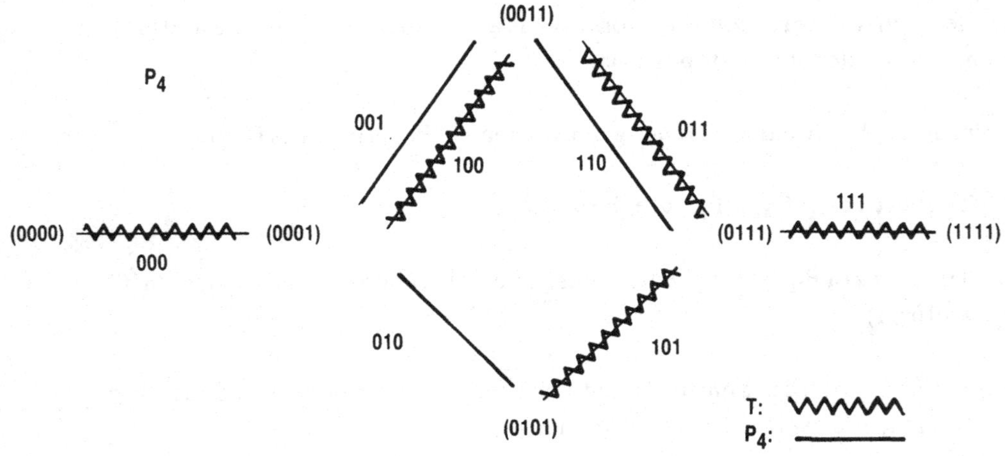

Figure 3. A Spanning Tree, T, of P_4

α	000	001	010	011	100	101	110	111
F(α)	1	0	0	1	1	1	0	1

Table 1. A Binary Truth Table

For the k-ary de Bruijn sequence case, the definitions of de Bruijn sequence, state, atom, link and cyclic NLFSR are the same as in the binary case, except that now $L = \{0,...,k-1\}$. The first complication arises in the definition of truth tables.

Definition: Let $L = \{0,...,k-1\}$ and let S_k be the permutation group on L. A **truth table** is a mapping $f: L^{n-1} \rightarrow S_k$.

Proposition: An NLFSR $M: L^n \rightarrow L^n$ with feedback function $F: L^n \rightarrow L$ is cyclic if and only if there is a truth table $f: L^{n-1} \rightarrow S_k$ such that $F(\alpha | x) = f(\alpha)(x)$ for all $\alpha \epsilon L^{n-1}$ and $x \epsilon L$.

For the proof of this and all subsequent results of this section, the reader is referred to Rabinowitz.[4]

Corollary: There is a natural one-to-one correspondence between truth tables and cyclic NLFSRs of span n over $L = \{0,...,k\text{-}1\}$.

In order to extend the spanning-tree result above to the k-ary case, we must first extend the definition of P_n. In the k-ary case, subgraphs correspond to classes of NLFSRs. As a result, shift registers must be categorized. This is the motivation for considering reduced truth tables.

Definition: A **partition** of L is a family of pairwise disjoint subsets of L with union L. We denote by B_k the set of partitions of L.

Definition: We denote by p: $S_k \rightarrow B_k$ the natural surjection. If $s \epsilon S_k$, then s has a canonical representation as a product of disjoint cycles. This specifies the partition p(s). For example,

$$p:(124)(03)(5) \rightarrow \langle \{1,2,4\},\{0,3\},\{5\} \rangle.$$

Definition: A **reduced truth table** is a function R: $L^{n-1} \rightarrow B_k$. The **reduction** of a truth table, F, is the reduced truth table, F_{red}, where $F_{red}(\alpha) = p(F(\alpha))$.

Definition: A truth table F is **distinguished** if the corresponding NLFSR outputs a de Bruijn sequence. A reduced truth table, R, is **distinguished** if every truth table with reduction R is distinguished.

Definition: A **hypergraph**, G, is a pair (V(G),E(G)) where V(G) is a finite set and E(G) is a finite family of (not necessarily distinct) subsets of V(G), each of which has cardinality at least two.

Subgraphs, trees, and spanning trees can be defined in the hypergraph case in a manner analogous to the graph case. A **subgraph** of G is a hypergraph that may be obtained from G by deleting, shrinking, and subdividing edges. In figure 4, H is a subgraph of G but K is not. A **tree** is a connected hypergraph that contains no cycles and has no two of its edges intersected in more than one vertex (figure 5). A **spanning tree** of G is a subgraph of G that is a tree and contains every vertex in at least one edge. These notions have been precisely defined elsewhere.[4,6]

Definition: The **link hypergraph** of span n over L = {0,...,k−1} denoted $P_{n,k}$, is given as follows:

(a) The **vertices** of $P_{n,k}$, $V(P_{n,k})$, are the atoms.

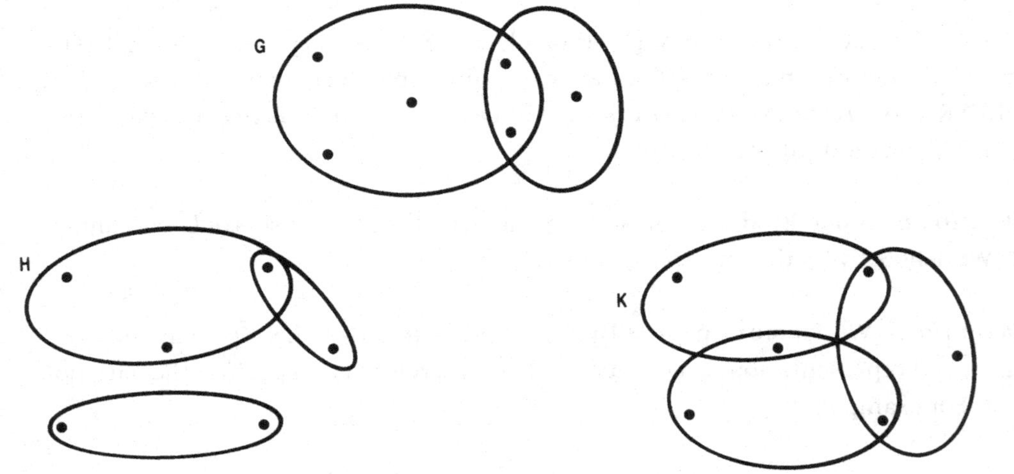

H is a subgraph of G; K is not a subgraph of G

Figure 4. Subgraphs of Hypergraphs

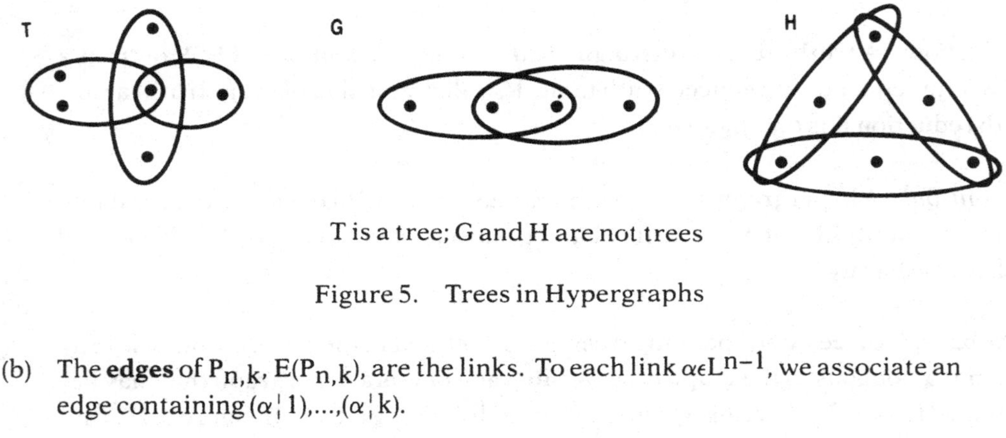

T is a tree; G and H are not trees

Figure 5. Trees in Hypergraphs

(b) The **edges** of $P_{n,k}$, $E(P_{n,k})$, are the links. To each link $\alpha \epsilon L^{n-1}$, we associate an edge containing $(\alpha|1),\ldots,(\alpha|k)$.

Proposition: There is a natural one-to-one correspondence between subgraphs of $P_{n,k}$ and reduced truth tables. More specifically, if R is a reduced truth table, then, for each $\alpha \epsilon L^{n-1}$, $R(\alpha)$ is a partition of L into a family of pairwise disjoint subsets, $E_1{}^{\alpha},\ldots,E_{r_\alpha}{}^{\alpha}$. Define a subgraph G of $P_{n,k}$ by letting E(G) consist of all edges of the form $E_i{}^{\alpha} = \{(\alpha|y_1),\ldots,(\alpha|y_m)\}$, where $\alpha \epsilon L^{n-1}$ and $\{y_1,\ldots,y_m\}$ is an element of $R(\alpha)$ with $m \geq 2$. This establishes the desired correspondence.

Definition: A subgraph of $P_{n,k}$ is a **d-subgraph** if its corresponding reduced truth table is distinguished.

Theorem: Every spanning tree of $P_{n,k}$ is a d-subgraph. In other words, if M is a cyclic NLFSR with truth table f such that f_{red} corresponds to a spanning tree of $P_{n,k}$, then M outputs a k-ary de Bruijn sequence of span n.

Example: Consider the spanning tree of $P_{2,3}$ in figure 6 which defines the reduced truth table R (table 2a). F (table 2b) is a truth table with reduction R. The output of the NLFSR with truth table F is the de Bruijn sequence with period 202210011.

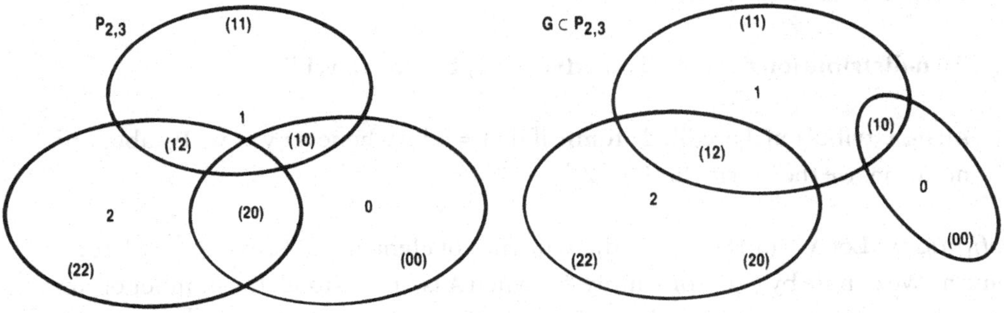

Figure 6. A Spanning Tree of $P_{2,3}$

(a)

a	R(a)
1	$\langle\{1,2,0\}\rangle$
2	$\langle\{1,2,0\}\rangle$
0	$\langle\{1,0\},\{2\}\rangle$

(b)

a	F(a)
1	(102)
2	(120)
0	(10)(2)

Table 2. A Reduced Truth Table and a Corresponding Truth Table

3. Atomic Sequences

In this section, we introduce a broad class of sequences of which the de Bruijn sequences are a special case. L is the finite alphabet L = $\{0,\ldots,k\text{-}1\}$, and k is some fixed integer, $k \geq 2$.

Definition:

(a) An **n-state** is an element of L^n.

(b) An **n-atom** is an element of $L^n/\sim$ where $v \sim w$ means that v is a cyclic shift of w. A_n (or $A_n(L)$) denotes the set of n-atoms.

(c) An **n-link** is an element of L^{n-1}.

Definition: Let $Z^+ = \{0,1,2,...\}$.

(a) An **n-distribution** is a function $d:L^n \to Z^+$, that is, an assignment to each n-state of a nonnegative integer.

(b) An **n-distribution** d is called **positive** if $d(v) \geq 1$ for all $v \epsilon L^n$.

(c) An n-distribution d is called **atomic** if $d(v) = d(w)$ whenever $v \sim w$. By abuse of notation, we then write $d:A^n \to Z^+$.

Definition: Let $A = (a_i)$ be a periodic sequence of elements of L, and let $s \epsilon L^n$ for some n. We denote by $\#_A(s)$, or simply $\#(s)$ when A is understood, the number of times that s appears in a single period of A. Thus, $\#_A$ is a mapping $\#_A:L_n \to Z^+$ for each A, that is, $\#_A$ is an n-distribution.

Example: If A is a de Bruijn sequence then $\#_A(s) = 1$ for all $s \epsilon L^n$, where n is the span of A.

Definition:

(a) A sequence A is **n-regular** if $\#_A:L^n \to Z^+$ is constant, i.e., if all n-tuples appear the same number of times in a single period of A.

(b) A sequence A is **n-atomic** if $\#_A$ is invariant under the action of cyclic shifting on L^n, that is, if $v \sim w$ implies $\#_A(v) = \#_A(w)$ for all $v,w \epsilon L^n$.

Example:

(a) Every periodic sequence is 1-atomic.

(b) If (a_i) has period m, then it is m-atomic.

(c) A de Bruijn sequence of span n is n-regular.

Lemma:

(a) If A is n-regular then A is n-atomic.

(b) If A is n-regular then A is r-regular when $r < n$.

Proof:

(a) follows immediately from the definitions. For (b) note that if every n-tuple appears m times in a period of A then every r-tuple appears $k^{n-r}m$ times.

Definition: An n-distribution d is **realizable** if there exists a sequence A such that $d = \#_A$.

Our aim is to characterize realizable atomic distributions and to provide an algorithm for the synthesis of realizing sequences.

4. Link Hypergraphs and Realizable Atomic Distributions

In this section, we extend the techniques of Section 2 to arbitrary atomic n-distributions. We associate to each such distribution a hypergraph, the spanning trees of which can be used to synthesize sequences that realize this distribution. The main difficulty lies in extending all the definitions and constructions. Once this is accomplished, previously established proofs[4] can be extended to this more general case. Accordingly, we content ourselves here with a careful description of the necessary generalizations and omit most proofs.

Definition: Let $L = \{0,1,...,k\text{-}1\}$ and let $I = \{1,...,r\}$ with $k \geq 2$ and $r \geq 1$. A **modified NLFSR** of span n over L with index set I (figure 7) is a quintuple {Q,M,F,G,H} satisfying the following properties:

(a) Q is a subset of $L^n \times I$.

(b) If $[v:j] \in Q$ where $v \in L^n$ and $j \in I$ then $[v:i] \in Q$ for $i < j$.

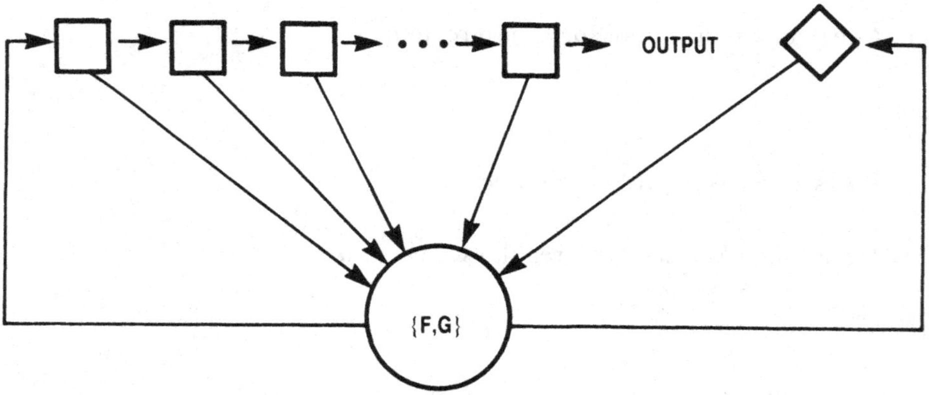

Figure 7. A Modified Nonlinear Feedback Shift Register

(c) $[v{:}r] \epsilon Q$ for some $v \epsilon L^n$.

(d) $M{:}Q \rightarrow Q$, $F{:}Q \rightarrow L$, $G{:}Q \rightarrow I$ and

$$M([v{:}j]) = M([v_{n-1},\ldots,v_0{:}j]) = [F([v{:}j]), v_{n-1},\ldots,v_1{:}G([v{:}j])]$$

(e) $H{:}Q \rightarrow L$, $H([v_{n-1},\ldots,v_0{:}j]) = v_0$.

We refer to Q as the **domain** of the shift register, to H as the **output function**, and to the pair {F,G} as the **feedback function**. By abuse of notation, M is identified with the shift register.

Definition: The **output sequence** of the modified NLFSR, M, initialized at $[v{:}j] \epsilon Q$ is the sequence $A = (a_i)$ where $a_i = H(M^{i-1}([v{:}j]))$. Note that $a_i = v_i$ for $i = 0,\ldots,n-1$.

Definition: Let $Q \subset L^n \times I$ be a domain and let $Q' \subset L^{n-1}$ be defined by $\alpha \epsilon Q'$ if and only if there exists an $x \epsilon L$ and $j \epsilon I$ such that $[\alpha \mid x{:}j] \epsilon Q$. For each $\alpha \epsilon Q'$, q_α denotes the set

$$q_\alpha = \{(x{:}j) \epsilon L \times I \mid [\alpha \mid x{:}j] \epsilon Q\}.$$

Let $S(q_\alpha)$ be the permutation group on q_α and let $B(q_\alpha)$ be the set of partitions of q_α.

(a) A **truth table** on Q is a function $T{:}Q' \rightarrow \cup S(q_\alpha)$ such that $T(\alpha) \epsilon S(q_\alpha)$ for all α.

(b) **A reduced truth table on Q** is a function $R:Q' \rightarrow \cup B(q_\alpha)$ such that $R(_\alpha) \epsilon B(q_\alpha)$ for all α.

(c) **Reduction** is as defined in section 2.

Definition: A modified NLFSR, M, is **cyclic** if $M:Q \rightarrow Q$ is a one-to-one mapping.

Proposition: A modified NLFSR, M, is cyclic if and only if there is a truth table $T:Q' \rightarrow \cup S(q_\alpha)$ such that the feedback function $\{F,G\}$ of M is given by

$$\{F,G\}([v_{n-1},...,v_0:j]) = T(v_{n-1},...,v_1)(v_0:j)$$

Given an atomic n-distribution, d, our aim is to construct a modified NLFSR whose output sequence realizes d.

Definition: Let d be an atomic n-distribution over L. We construct a hypergraph P_d as follows:

(a) $V(P_d) = \{[(v):j] \mid (v) \epsilon A_n, 1 \leq j \leq d(v)\}$. This is well-defined because d is atomic.

(b) For $\alpha \epsilon L^{n-1}$, let $e_\alpha = \{[(v):j] \epsilon V(P_d) \mid v \sim \alpha \mid x$ for some $x \epsilon L\}$.$E(P_d) = \{e_\alpha \mid e_\alpha$ has cardinality $\geq 2\}$. P_d is the **link hypergraph associated to d.**

Proposition: Let $d: L^n \rightarrow Z^+$ be an atomic n-distribution and let r be defined by $r = \max d(v)$. Define a domain Q by

$$Q = \{[v:j] \epsilon L^n \times I \mid 1 \leq j \leq d(v)\}$$

where $I = \{1,...,r\}$. Then there is a natural one-to-one correspondence between subgraphs of P_d and reduced truth tables on Q.

Proof: Let R be a reduced truth table on Q, and let α be an element of Q′. Then $R(\alpha) \epsilon B(q_\alpha)$, that is, $R(\alpha)$ is a partition of q_α. Let U be an element of this partition of cardinality ≥ 2, say $U = \{(x_i:j_i) \mid i=1,...,m\}$. Then $[\alpha \mid x_i:j_i] \epsilon Q$ so $j_i \leq d(\alpha \mid x_i)$ for all i. Let $E_U \subset V(P_d)$ be defined by $E_U = \{(\alpha \mid x_i):j_i] \mid i = 1,...,m\}$. Then $E_U \subset e_\alpha$ and by continuing this process for all such U and then for all α, we obtain a subgraph of P_d, Q.E.D.

Theorem: An atomic n-distribution, d, is realizable if and only if P_d is connected. More specifically, suppose P_d is connected and S is a spanning tree of P_d. Let R be the reduced truth table corresponding to the subgraph S of P_d, and let T be a truth table with reduction R. Finally, let M be the cyclic modified NLFSR with truth table T. The output sequence of M has distribution d.

Corollary: Every positive atomic n-distribution is realizable.

Proof: If d is positive, then P_d is connected.

Example: Let L = {0,1}, let n = 2 and let d be given by d(00) = 1, d(11) = 3, and d(01) = d(10) = 2. Consider the spanning tree of P_d in figure 8, which defines the reduced truth table R (table 3a). F (table 3b) is a truth table with reduction R. The output sequence of the modified NLFSR with truth table R is the sequence with period 10111100 realizing the distribution d.

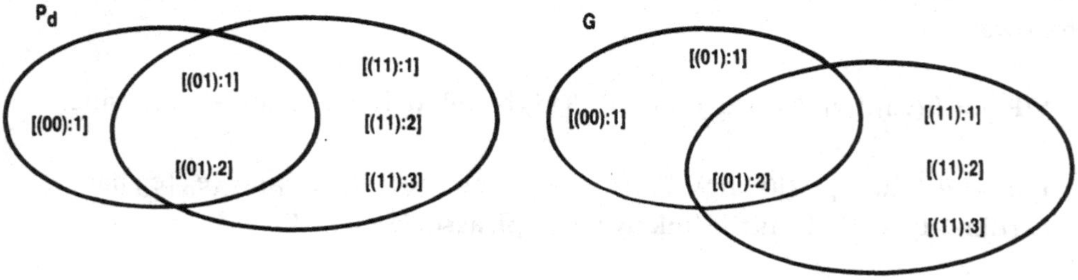

Figure 8. A Spanning Tree of P_d

(a)

a	R(a)
0	⟨{(0:1), (1:1), (1:2)}⟩
1	⟨{(0:1)},{(0:2), (1:1), (1:2), (1:3)}⟩

(b)

a	F(a)
0	((0:1) (1:2) (1:1))
1	((0:2) (1:2) (1:3) (1:1))

Table 3. A Reduced Truth Table and a Corresponding Truth Table

References

1. de Bruijn, N. G., A combinatorial problem, *Pros. Neder. Akad. Wetensch,* 49, 1946.

2. Fredricksen, H., A survey of full length nonlinear shift register cycle algorithms, *SIAM Review,* 24, 195, 1982.

3. von Aardenne-Ehrenfest, T. and de Bruijn, N.G., Circuits and trees in ordered linear graphs, *Simon Steven,* 28, 203, 1951.

4. Rabinowitz, J. H., de Bruijn sequences and hypergraphs over finite alphabets, *Proceedings of the Thirteenth Southeastern Conference on Combinatorics, Graph Theory, and Computing, Congressus Numerantium,* Utilitas Math., Winnipeg, to appear.

5. Golomb, S. W., *Shift Register Sequences,* Aegean Park Press, Laguna Hills, 1982.

6. Berge, C., *Graphs and Hypergraphs,* North-Holland, Amsterdam-London, 1973.

LIST OF CONTRIBUTORS

Toby BERGER and Nader MEHRAVARI, School of Electrical Engineering, Cornell University, Ithaca, New York 14853 (USA).

Thomas BETH, Inst. f. Mathematische Maschinen u. Datenverarbeitung, Universität Erlangen-Nürnberg, Martensstrasse 3, 8520 Erlangen (Western Germany).

Richard E. BLAHUT, IBM Corporation, Bodle Hill Road, Owego, N.Y. 13827 (USA).

Marc DAVIO and Jean-Marie GOETHALS, Philips Research Laboratories, Avenue Van Becelaere 2, Box 8, 1170 Bruxelles (Belgium).

Anthony EPHREMIDES, Electrical Engineering Department, University of Maryland, College Park, Maryland 20742 (USA).

Sami HARARI, Université de Picardie, U.E.R. de Mathematiques, Rue Saint Leu, 80039 Amiens, Cedex (France).

Robert McELIECE, California Institute of Technology, 116-81, Pasadena, California 91125 (USA).

Joshua H. RABINOWITZ, The Mitre Corporation, Mail Stop E095, Bedford, Mass. 01730 (USA).

Andrea SGARRO, Istituto di Elettrotecnica ed Elettronica, Università di Trieste, Via A. Valerio 10, Trieste (Italy).

Printed in the United States
By Bookmasters